Wood as Building and Hobby Material

Wood as Building and Hobby Material

How to Use Lumber, Wood-Base Panels, and Roundwood Wisely in Construction, for Furniture, and as Fuel

HANS KUBLER

Associate Professor
University of Wisconsin
Madison, Wisconsin

A WILEY-INTERSCIENCE PUBLICATION

JOHN WILEY & SONS
New York • Chichester • Brisbane • Toronto

Library of Congress Cataloging in Publication Data:

Kubler, Hans.
Wood as building and hobby material.

"A Wiley-Interscience publication."
Bibliography: p.
Includes index.
1. Wood. 2. Building, Wooden. 3. Woodwork.
4. Wood as fuel. I. Title.

TA419.K73 620.1'2 80-13380
ISBN 0-471-05390-2

Printed in the United States of America

10 9 8 7 6 5 4 3 2 1

To the Memory of my Late Mother

Preface

This book does **not** tell you *all you want to know about wood;* it offers something better: a small quantity of useful concepts which occupy little space in your head, but—when applied with common sense—are very prolific and serve you better than a myriad of details. The book leads straight into a vast field of facts, explaining wood's essential features, on whose basis you will understand the wide variety of wood properties and the wood-base panel products.

I wrote this text for builders and hobbyists, for architects and artists, construction administrators and wood-burners, craftsmen and civil engineers, as well as teachers of industrial arts and everybody who has an interest in this fascinating, universal material wood. Foresters may find the book of interest in their work of producing timber. For wood technology students it should be suitable for introductory material.

The modern wood-base panel products—plywood, particleboard, hardboard, and insulating board—comprise a small fraction of the outline but a significant portion of the book. Many sections deal with the panels indirectly. I selected the bibliography for readers who want additional information. Many of the listed U.S. Forest Products Laboratory (FPL) publications are available free of charge from the FPL (Box 5130, Madison, Wis. 53705). Readers who want technical information from the industry may request the booklet *Lumber and Wood Products Literature* from the National Forest Products Association (1619 Massachusetts Ave., N. W., Washington, D. C. 20036). The booklet lists hundreds of leaflets, brochures, books, sets of slides, and films, together with the address of the issuing association.

The text evolved from a course taught at the University of Wisconsin for construction technology majors and many other students who—having some interest or another in wood—took the course as an elective. The varying backgrounds of the audience forced me to assume little previous knowledge, avoid technical language, use many clear illustrations, and restrict the teaching to subjects relevant for the majority.

Through the years guest lecturers from the staff of the FPL contributed much to the success of the course and left their mark in this book; they were: R. A. Bendtsen, J. M. Black, E. H. Bulgrin, H. W. Eickner, G. R. Esenther, W. E. Eslyn, W. C. Feist, R. H. Gillespie, L. R. Gjovik, R. E. Jones, J. F. Lutz, J. D.

McNatt, R. C. Moody, and especially R. C. Koeppen. Among other contributions I recognize the excellent editorial assistance of Ann White and my former wife Sally, the art work of Jean Holland, and Pauline Meudt's diligent typing.

HANS KUBLER

Madison, Wisconsin
June 1980

Contents

Wood as Building and Hobby Material

1

Tree-Related Properties and Structure of Wood

Human ingenuity can never devise anything more simple and more beautiful, and more to the point than nature does.

Leonardo da Vinci

Wood is a wonderful material for building a tree.

Wisecrack

Living things abound in miraculous features, but trees in a way top all other organisms: no other plants and no animals attain such height and such age. This outstanding performance requires proper functioning of all parts, especially of wood, a tree's most characteristic component. Wood is designed for the needs of the tree, it serves the tree, and is structured accordingly—facts that help us understand wood and wood properties.

Though wood has the same functions in all kinds of trees, its appearance and structure differ so much from kind to kind that the species can almost always be identified. The Appendix deals with identifying 30 important commercial woods, while this chapter covers structural features in connection with other properties.

1.1. NAMING AND CLASSIFICATION OF WOOD AND TREES

Wood is not always named after the tree from which it originates, and wood of a certain name may come from more than one tree species. *Hard maple* lumber, for example, is sawn from the *sugar maple* or from the *black maple* tree. Conversely, lumber from one and the same kind of tree may appear under different names in different regions, even in the same city. Clever dealers invent names to make it difficult for competitors to locate the timber, also to give that wood a favorable image (*African mahogany* was named in this way; it differs from Central American *mahogany* and is not simply Africa-grown *mahogany*).

Two standards bring order into this confused state: *Product Standard PS 20–70* (*American Softwood Lumber Standard*), discussed in Section 5.1, and *ASTM D 1165* (*Standard Nomenclature of Domestic Hardwoods and Softwoods*) of the American Society for Testing and Materials (see the Bibliography at the end of the book). *PS 20-70* lists certain *commercial names for lumber* for 54 softwood tree species. Standard *D 1165* codes the commercial practice and retains more than one name for many of the nearly 150 commercially important tree species growing in the United States and Canada. The names in Table 1.1 and in other chapters of this book comply with the two standards. The *official common tree names* listed in Table 1.1 are used by the U. S. Forest Service and have found general acceptance. Some *tree names* differ from *commercial names for lumber* in spelling. I prefer the latter in this book.

The international *botanical names* or *scientific names of plants* in Latin follow simple concepts and tell a good deal even to readers who do not know Latin. The name of the tree species sugar maple, *Acer saccharum* Marsh., serves as an example:

Acer	Stands for the *genus*, it is the generic name of maples
saccharum	Is the specific name, in this case meaning *sugar*, referring to the tree's sweet sap
Marsh.	Indicates the surname of the author who named and described the species first in a scientific publication; this third part is often omitted

The various species (*species* being a word that refers to both singular and plural) of the genus *Acer* have certain features in common, but this is of limited benefit in learning about the wood because botanists classify mainly by reproductive organs—in rare cases placing several species of trees and even other plants into the same genus. Trees with similar reproductive organs do not always have similar wood. The situation is the same with leaves, as this example illustrates: most *Acer* species have simple palmately lobed leaves, while the leaves of boxelder (*Acer negundo*) are odd-pinnate and consist of several leaflets. Boxelder nevertheless belongs to the genus *Acer* because it bears the characteristic V-shaped winged fruits of the maples.

In the literature appear botanical names like *Acer* spp. and *Acer* spec. *Spec.* is the abbreviation of the Latin word *specificus*, meaning one certain

Table 1.1. Nomenclature for Some[a] Domestic Hardwoods and Softwoods

Commercial Name for Lumber	Official Common Tree Name[b]	Botanical Name
	HARDWOODS[c]	
Alder, red	Red alder	*Alnus rubra*
Ash, white[d]	White ash	*Fraxinus americana*
Aspen[d]	Quaking aspen[e]	*Populus tremuloides*
Basswood	American basswood	*Tilia americana*
Beech	American beech	*Fagus grandifolia*
Birch[d]	Sweet birch	*Betula lenta*
	Yellow birch	*B. alleghaniensis*
Butternut	Butternut	*Juglans cinera*
Cherry	Black cherry	*Prunus serotina*
Cottonwood[d]	Eastern cottonwood	*Populus deltoides*
Elm:		
Rock[d]	Rock elm	*Ulmus thomasii*
Soft[d]	American elm	*U. americana*
Gum	Sweetgum[e]	*Liquidambar styraciflua*
Hickory[d]	Shagbark hickory	*Carya ovata*
Locust[d]	Honeylocust	*Gleditsia triacanthos*
Magnolia[d]	Southern magnolia	*Magnolia grandifolia*
Maple, hard[d]	Sugar maple	*Acer saccharum*
Oak:		
red[d]	Northern red oak	*Quercus rubra*
white[d]	Live oak	*Q. virginiana*
	White oak	*Q. alba*
Sassafras	Sassafras	*Sassafras albidum*
Sycamore	American sycamore	*Platanus occidentalis*
Tupelo[d]	Black tupelo	*Nyssa sylvatica*
Walnut	Black walnut	*Juglans nigra*
Yellow poplar	Yellow poplar	*Liriodendron tulipifera*
	SOFTWOODS[f]	
Cedar, western red	Western redcedar	*Thuja plicata*
Cypress	Baldcypress[e]	*Taxodium distichum*
Douglas fir	Douglas-fir	*Pseudotsuga menziesii*
Fir, white[d]	White fir	*Abies concolor*
Hemlock, west coast	Western hemlock	*Tsuga heterophylla*
Larch, western	Western larch	*Larix occidentalis*

Table 1.1. (*Continued*)

Commercial Name for Lumber	Official Common Tree Name[b]	Botanical Name
Pine:		
Northern white	Eastern white pine	Pinus strobus
Ponderosa	Ponderosa pine	P. ponderosa
Southern yellow	Loblolly	P. taeda
	Longleaf[e]	P. palustris
	Shortleaf	P. echinata
Redwood	Redwood	Sequoia sempervirens
Spruce, Engelmann	Engelmann spruce	Picea engelmannii

[a] Selected for economic significance and interesting properties.
[b] The tree names conform to the Check List of Native Trees of the United States (including Alaska) by E. L. Little, U.S.D.A. Agr. Handb. 41, Washington, D.C., 1953.
[c] Source: Standard Nomenclature of Domestic Hardwoods and Softwoods, ASTM D 1165-76.
[d] The commercial name applies also to lumber from other tree species.
[e] Lumber of the species is marketed also under other commercial names.
[f] Source: American Softwood Lumber Standard, PS 20-70.

species. *Spp.* stands for the species in the plural and is used for a number of species of the genus.

Botanists organized similar genera (plural of *genus*) into a *family*, similar families into an *order*, similar orders into a *class*, and similar classes into a *phylum*. Wood technology rarely benefits from classification on these levels, since members of one family of wood may resemble those of another family more than its own. On one important level the classification is consistent: in each of the two classes of *gymnosperms* and *angiosperms*, all woods share certain features. The world's trees, in fact, belong to these two classes, which together with the class of ferns, constitute the most highly developed phylum of *vascular plants*, meaning *plants with conducting tissue.* All common plants are of the vascular type.

Gymnosperms (from Greek *gymnos*, naked, and *sperma*, a seed), produce bare seeds that are not surrounded by floral parts. Within the gymnosperm class all important trees appear in the order of *conifers,* meaning *cone bearers,* using woody cones to carry their seeds. Yew and several other conifers do not bear cones, but the cone is such a conspicuous feature of most conifers that the order has been named after it. Conifers dominate many forests in the temperate zone but do not grow in tropical climates. Their leaves are either needle-like as on pines, spruces, and firs, or scale-like as on cedars. Most conifers retain their leaves in winter as *evergreens*; larches and baldcypress represent exceptions. We also call coniferous trees *softwoods,* as explained below.

Seeds of angiosperms (from Greek *angeion*, capsule) are enclosed in pulpy tissue, forming a fruit as in apples, cherries, nuts, mulberries, and strawberries, which animals eat and inadvertently scatter. Most angiosperms do not qualify as trees or shrubs, are rather *herbaceous* plants, lacking woody tissue. Angiosperm trees are known as *broad-leaved* and as *hardwoods*. In zones with cold winters or severe dry seasons they are *deciduous,* which means they *shed leaves at the end of the growing season.*

The terms *softwood* and *hardwood* may mislead. *Hard* means *not easily dented* and the majority of hardwood trees have relatively hard wood, but some, such as the various kinds of aspen, the cottonwoods, and yellow poplar, are softer than most softwoods. Balsa, by far the softest of all, is a hardwood from tropical forests in South America. Similarly, wood of most softwood trees is relatively soft but yew and a few other softwoods exceed most hardwoods in hardness. The terms *coniferous wood* and *angiosperm wood* are more accurate than *softwood* and *hardwood.* Interestingly enough the Germans, Russians, and Spaniards use names like those in English, whereas the French use *resinous wood* for *softwood.* In old times, when in the temperate zone man utilized almost exclusively the hard species of broad-leaved trees and soft species of conifers, *softwood* and *hardwood* were appropriate terms. Wood hardness is crucial only in such uses as flooring, but it represents a very obvious property, is easy to evaluate, and is a good indicator of weight, strength, shrinkage, workability, and other characteristics as explained in Section 4.1.

1.2. RELATION OF FUNCTIONS, STRUCTURE, AND PROPERTIES

Wood in the tree trunk supports the branch and leaf structure, conducts water solutions (*sap*) to the crown, and stores the carbohydrates, which are synthesized in leaves to flow as dissolved sugars down in the inner bark and to migrate from the bark into the wood. These three functions determine wood's essential features.

Cell Structure and Structural Terms

Engineers have learned to support loads with a minimum of material, using hollow steel cylinders. In the form of pipes these are also well suited for conducting liquid. Similarly a tree trunk consists mainly of tubular cells, oriented *longitudinally,* parallel to the trunk axis.

In softwoods, it is the *tracheid* cells that provide mechanical support and conduct sap. Tracheids are a particular type of *fiber*; they measure 1/16 to 1/2 in. in length but only 1/100 of that in diameter; their tapered ends overlap for good connection. Tracheids represent 90% of softwood tissues (Fig. 1.1*a*).

Hardwoods evolved long after softwoods, and form specialized tissue with wide sap-conducting *vessels* (discussed below), as well as fibers which support mechanically. On the average, hardwood fibers occupy one half of hardwood tissue (Fig. 1.1*b*). Being shorter than tracheids they were considered unsuitable for making paper, which is a mat of interfelted fibers, but engineers learned to make good paper also from hardwood.

The fiber direction generally coincides with the longitudinal direction. *Fiber* in the broad sense refers to both hardwood fibers and tracheids. The term *grain* includes fibers in the broad sense and also vessels—hence the term *along the grain* for *fiber direction. Perpendicular to the grain* or *across the grain* means *perpendicular to the fibers*—horizontal to the trunk of the standing tree—another term for this direction being *transverse.*

The orientation of the cells affects the appearance and many other properties of wood surfaces. Cuts perpendicular to the fiber direction expose *transverse sections,* also called *cross sections* (Fig. 1.2), *end-grain surfaces,* and *end faces.* These are naturally much rougher and more porous than *lateral sections* or *side-grain surfaces.*

The vessels or *pores* extend in fiber direction from the tips of the roots all the way up to the leaves. In oaks (Fig. 1.3B) and many other species the vessels are wide enough to be seen with the unaided eye as holes at transverse sections, and as grooves at lateral sections, provided the surface has been planed, sanded, or cut smooth with a sharp knife (on sawn surfaces torn-wood fragments obscure the pores). *Open-grained* wood features large pores, but *close-grained* does not mean *small pores* (Section 1.3). Vessel pores are not to be confused with the resin canals, explained in the Appendix.

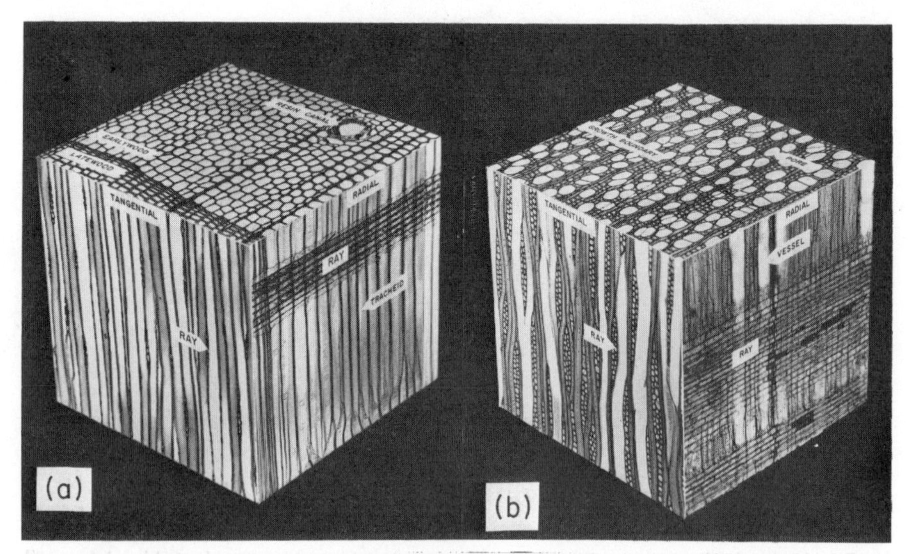

Figure 1.1. Microphotographic model of softwood (*a*) and diffuse-porous hardwood (*b*). (From Kollmann and Côté, 1968.)

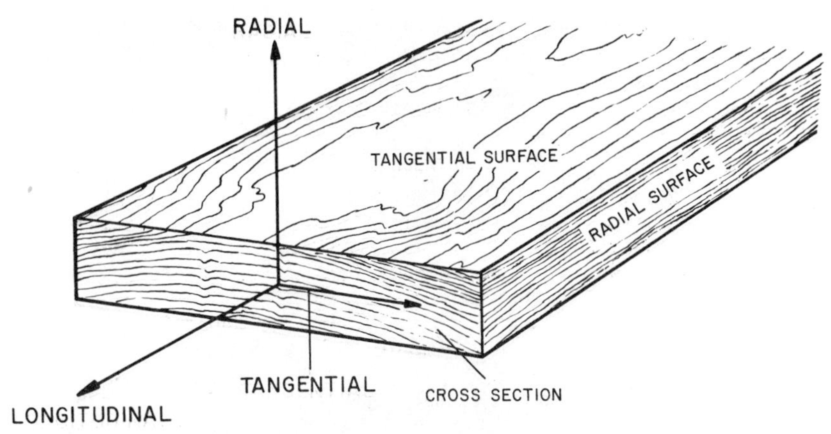

Figure 1.2. The three principal axes and surfaces of wood with respect to fiber direction and annual rings. (Photograph courtesy of U.S. Forest Products Laboratory.)

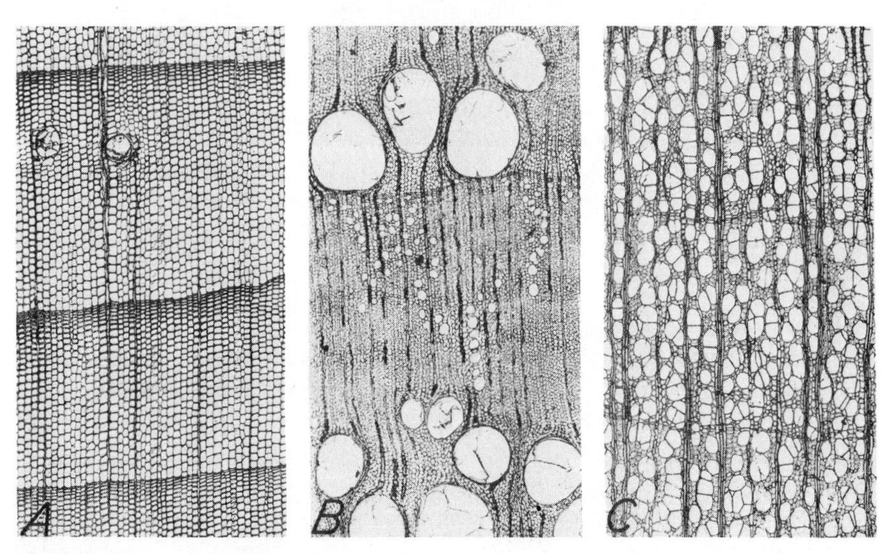

Figure 1.3. Cross section microphotographs of northern white pine (*A*), white oak (*B*), and hard maple (*C*) as examples of softwood, ring-porous hardwood, and diffuse-porous hardwood. (Photograph courtesy of U.S. Forest Products Laboratory.)

Each vessel consists of numerous *vessel-member* cells, stacked one above the other to form a continuous pipe. In species with wide pores the vessel members are shorter than wide, and resemble drums or barrels rather than tubes. The vessel-member cavities are connected either by several relatively large openings in the dividing plate, or by one single opening in the rim-like plate rudiment (Fig. 1.4). The openings determine the rate of sap flow and how permeable the wood is in the fiber direction.

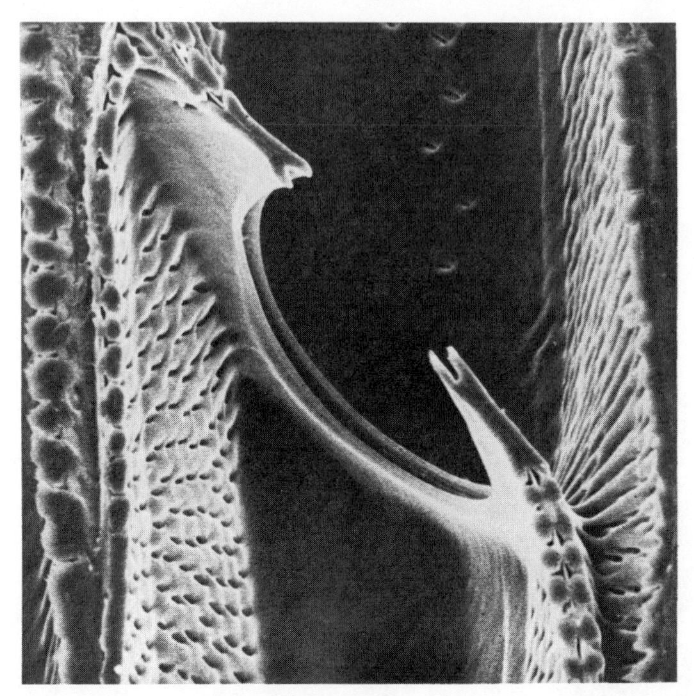

Figure 1.4. Scanning electron micrograph of the perforation plate between two vessel members in New Zealand honeysuckle. (From Meylan and Butterfield, Chapman & Hall, Ltd., 1972.)

Figure 1.1 shows *rays* for conducting sugar solutions from the bark into the wood, and for storing the converted sugars in the form of starch as food reserve. Rays consist of transverse cells, most of which are the brick-shaped *parenchyma* type with thin walls. All rays begin in the bark, extending transversely toward the trunk's center, and ending somewhere between bark and center (Fig. 1.5). Most hardwood rays are sufficiently large to be seen with the unaided eye and contribute to the characteristic appearance of the species. Cross sections show edges of the ray bands in the form of lines, while at lateral sections each ray looks like a band, a spindle, or a hybrid of band and spindle, depending on the cut (Fig. 1.6).

Structure-Related Properties

The orientation of fibers makes wood many times stronger longitudinally than transversely, as is necessary to carry the crown of the tree. The ratio of longitudinal to transverse strength averages 10:1, but it runs as high as 50:1 in tensile strength of some species. Wood permeability depends on this fiber direction almost as much as does strength, reflecting the fact that much more water flows upward in the trunk than sugar solution migrates transversely. Wood dries

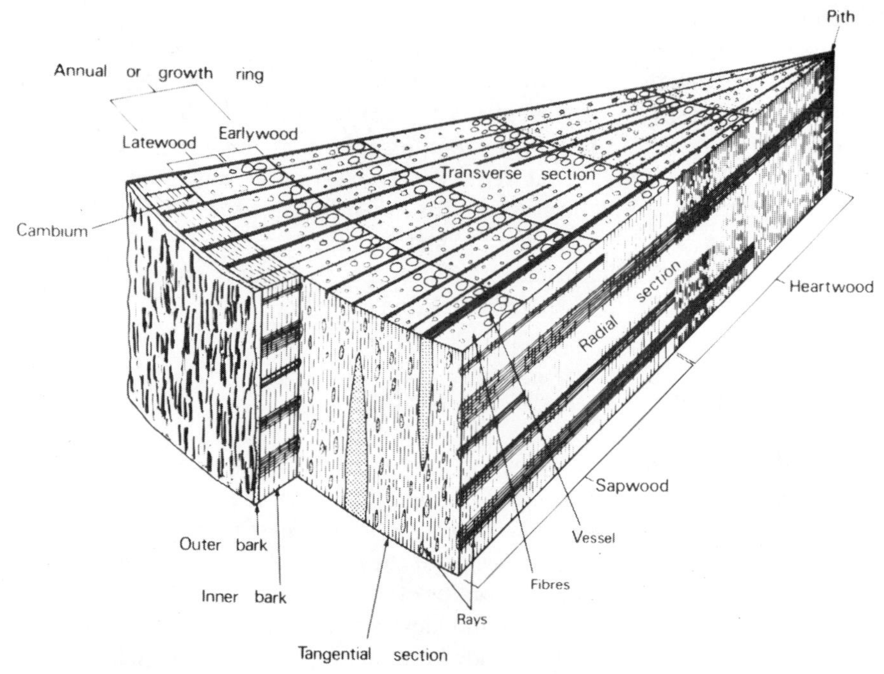

Figure 1.5. Diagrammatic wedge section of five year old oak trunk. (Crown copyright; photograph courtesy of Princes Risborough Laboratory of the Building Research Establishment, England.)

Figure 1.6. Appearance of rays on the radial section of hard maple.

therefore about 10 times faster at end-grain surfaces, and absorbs much more preservatives, glue, and finish there, than on lateral surfaces.

Wood is distinctly *anisotropic,* meaning its *properties vary with the direction.* The anisotropy shows spectacularly in *splitting*, when fibers separate in their direction ahead of the splitting tool, forming a gap between tool edge and the material (Fig. 1.7). Splitting requires less energy than any other method of wood disintegration and is applied where uneven rough surfaces are acceptable. Wood splits also when it should not: when nailed and screwed, during planing (Section 11.2), and in veneer cutting (Section 5.4). It splits most easily in the plane of ray bands of thin-walled parenchyma cells (Fig. 1.8).

1.3. TREE GROWTH AND ANNUAL RINGS

Trees grow lengthwise and in thickness. Growth in length is restricted to the tips of stems, branches, and roots. Five-in. high initials carved into a tree at breast height, for example, remain 5-in. high and stay at breast height no matter how much the tree grows. The tissue formed at the tips appears in wood as *pith,* the small soft core occurring near the center of a tree trunk, branch, or root. Pith tissue consists of parenchyma cells and lacks strength. In many species the pith is too small to be visible for the unaided eye; in others it can be seen at the central lateral surface as a fine dark line or narrow band, at the transverse surface as a dot. Pith downgrades lumber, for reasons explained in Section 3.7.

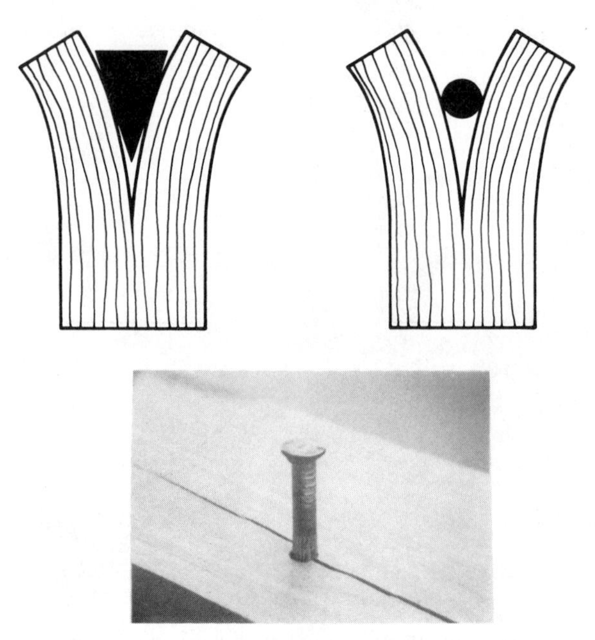

Figure 1.7. Splitting wood by wedge and by nail.

Figure 1.8. Wood splits along the grain but not across. (Photograph courtesy of R. C. Koeppen.)

Growth in Thickness

Between wood and bark a microscopically thin cell sheath extends from the tips of the roots to the tips of the branchlets, enclosing all the wood (Fig. 1.9). *Mother cells* of this *cambial layer* or *cambium* divide and produce new cells: wood cells toward the inside and bark cells to the outside. The new cells push the old bark outward, in this way increasing the trunk diameter.

New wood fibers and vessel members grow, mature, and die within a few weeks. Parenchyma cells live much longer; some stay alive for many years, but being encased in a skeleton of hard fibers and vessels they have no place to grow, except as tyloses (Section 1.5). Wood growth is restricted to the cambial zone.

While dividing and growing, the soft cambial cells provide weak bonds between wood and bark. In this state the bark easily peels off, separating from the tree at the cambial layer, a part of which remains attached to the bark while the other part sticks to wood. Tough dormant and dead cambial layers bond fairly well, so that debarking of live trees in winter and of sound dead trees in summer requires much more force than when the cambium is active; but even then the bond is weaker than the cohesion within bark itself and within wood. For effects of particular drying and storage conditions on adhesion of bark see Section 9.5.

Annual Rings

Wood cells consist of cell walls and empty cell cavities. In many tree species the *wall-to-cavity* ratio changes during the growing season. First-formed *earlywood* or *springwood* has relatively thin walls and large cavities and is more porous

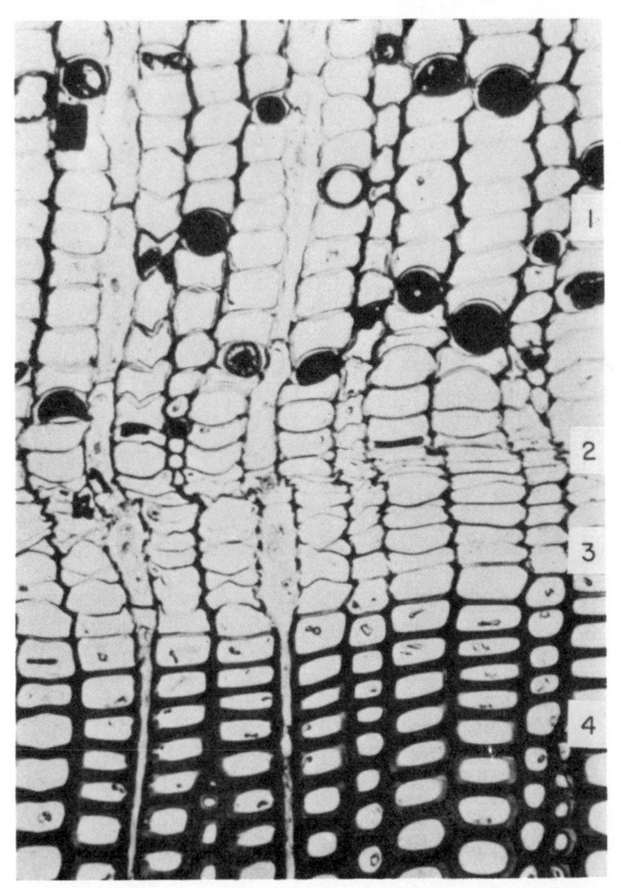

Figure 1.9. Microphotograph cross section of the active cambial layer (2) of pine lying between bark (1) and the immature tracheids (3) and wood (4). (Photograph courtesy of Forest Research Institute, Dehra Dun, India.)

than *latewood* or *summerwood* (Fig. 1.3*B*). At tree cross sections, earlywood and latewood appear as circular growth layers known as *annual rings*. At side-grain surfaces the cylindrical layers naturally show in various other shapes depending on the plane of cut (Fig. 1.10), but are called *rings* nevertheless. Annual rings together with rays and knots give surfaces particular *figure*.

The *wall-to-cavity* ratio of some species gradually changes throughout the ring so that no definite border between earlywood and latewood shows (Fig. 1.3*A*). Other species feature abrupt transition (Fig. 1.10), which when combined with relatively wide rings, gives lumber like southern yellow pine a coarse appearance and makes it hard to work.

In *ring-porous* hardwoods—such as most oaks, elm, ash, and hickory—earlywood vessels are many times wider than latewood vessels and form distinct rings or narrow bands (Fig. 1.3*B*). By contrast, in the *diffuse-porous* species—

Figure 1.10. Sector of southern yellow pine trunk showing transverse (X), radial (R), and tangential (T) surfaces and annual rings with light-colored earlywood and dark-colored latewood. (Photograph courtesy of Core, Côté, and Day, 1979.)

maple, yellow poplar, and many others—the vessels are of uniform size throughout the annual ring, the ring being marked only by a narrow line of thick-walled fibers and/or parenchyma cells at the very edge (Figs. 1.1*b* and 1.3C).

In tropical forests with uniform rainfall, trees grow continuously and not in annual rings. Some like mahogany produce narrow lines of parenchyma after various periods; many change the orientation of fibers and vessels by several degrees in periods of a few years, in this way giving the appearance of growth layers at some side-grain surfaces (see *interlocked grain* in Section 1.6). Distinct rings tend to enhance wood beauty, but in other respects people prefer a uniform structure (Sections 5.4, 11.2, and 14.3).

Ring-Related Terms

The circular form of growth layers at cross sections is the reason for two important designations: *radial* for the direction of the circle radii of the rays, and *tangential* meaning *along the tangent to the annual ring* (Fig. 1.2). Transverse directions that lie in between *radial* and *tangential* are called *intermediate*, but the term is uncommon; instead we use *radial* and *tangential,* whichever comes closest. *Tangential* stands for all transverse directions closer to the rings than to the rays, while other transverse directions are *radial.*

The terms *radial* and *tangential* designate not only transverse directions but also side-grain surfaces. Cuts approximately parallel to the growth layers are *tangential*; those closer to the planes of ray bands are *radial* (Fig. 1.2).

In the above-mentioned designation *along the grain*, the term *grain* denotes fibers and vessels in the visual sense of tiny particles. Unfortunately *grain* refers also to annual rings. *Close grain* means *narrow rings*, in contrast to the *coarse grain* of fast-growing trees with conspicuous rings.

In *edge-grained* lumber, rings form an angle of 45 to 90° with the surface, whereas in *flat-grained* lumber the angle is 0 to 45° (Fig. 1.11). Later chapters show that for most purposes edge-grained lumber is better. Sawmills therefore take special efforts to convert valuable hardwood logs into lumber of this kind, cutting the stem first into quarters to be *quarter-sawn*, in contrast to *plain-sawed*. For lumber consumers, *edge-grained* means *quartersawn* and *flat-grained* means *plainsawed*.

1.4. HOW HEARTWOOD AND SAPWOOD DIFFER

Cross sections of tree trunks typically show *bark*, a pale zone of *sapwood* surrounding dark *heartwood*, and the *pith* dot in the center (Fig. 1.12). Bark has little commercial value and does not fall into the scope of this book except as fuel (Section 9.5), while the pith has been considered in the preceding section. This section concerns the trunk's main parts; sapwood and heartwood. Laymen inaccurately assume that heartwood is stronger, harder, and shrinks less than sapwood. The difference between the two is more complex than that. On the basis of heartwood formation we shall understand how the two really differ.

Transition From Sapwood to Heartwood

In young trees all wood is sapwood. As the years pass on, some sapwood becomes heartwood, the transition starting near the pith and progressively seizing

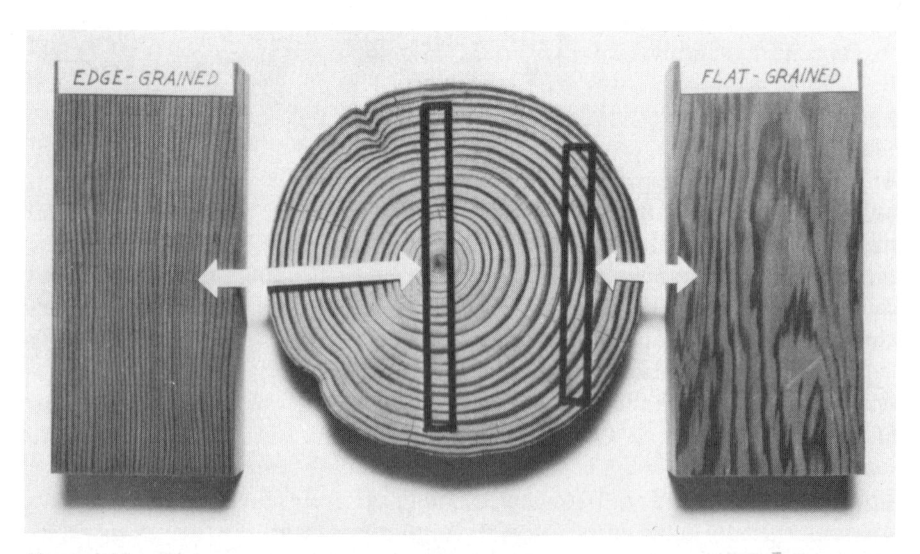

Figure 1.11. Edge-grained and flat-grained lumber in relation to the tree trunk. (Photograph courtesy of W. C. Feist.)

Figure 1.12. Cross section of bur oak showing pith, dark heartwood inside pale sapwood, and bark.

rings farther out, while new sapwood continues to grow in the cambial zone so that the sapwood zone approximately retains its width. Wood in the transition line ceases to conduct sap and gives up its food reserves for good: *heartwood* is defined as *wood which no longer participates in the life processes of the tree.* The transition involves death of the last surviving parenchyma cells; the detectable criterion of heartwood is absence of living cells.

In some tree species like the true firs (to which Douglas-fir does not belong), the transition process is just that simple. In these, heartwood looks like sapwood, the one difference being that stored food is absent in the heartwood. Many other species accumulate tannins, oils, and other *heartwood deposits,* which are either formed along the transition line or pass in from young sapwood and may originate in the leaves. Many deposits are of a particular color: red-tinged brown in red oak, red in rosewood, and nearly black in ebony, for some examples. In Alaska cedar and Port Orford cedar the deposits have the same color as sapwood, so that heartwood and sapwood look alike though differing chemically.

Two other parts of heartwood formation—growth of tyloses and pit aspiration—are explained in the next section, while changes in moisture content follow in Section 3.4.

Durability, Strength, and Dimensional Stability

Some heartwood deposits are poisonous to wood-damaging organisms and preserve the heartwood of redwood, black locust, several kinds of oak, and of many other species. We may surmise that trees produce the poisons to protect the heartwood, which—though no longer participating in life processes—still helps carry the tree crown. Sapwood needs no protection since the sap suffocates invading organisms; sap poison, if it existed, might even harm living tree cells. According to another opinion, heartwood deposits are waste products, and heartwood together with bark serve as the tree's dump, a view which is supported by the fact that in some species the heartwood deposits are nonpoisonous to wood-damaging organisms.

Hence a dark-colored heartwood is not a reliable indication of durability, nor does sapwood-like appearance necessarily speak for sapwood-like susceptibility, even though, more often than not a darkly colored heartwood is more durable than sapwood. The different modes of heartwood formation are illustrated in Fig. 1.13. Section 10.9 deals with durability from wider view points.

Some heartwood deposits remaining in cell cavities are oils that repel water; some are hard minerals that dull tools (Section 11.3). Other deposits infiltrate cell walls, and strictly speaking, do influence strength and moisture relations, but generally are too few for significant effects. The deposits reach substantial proportions in the order of 10% only in a few species such as redwood, western red cedar, and black locust, whose heartwood is slightly stronger and less hygroscopic than sapwood (Section 3.2). The erroneous view of heartwood being generally very strong results from comparing rotten sapwood with sound heartwood.

1.5. RISE OF SAP IN TREES AND PERMEABILITY OF WOOD

In trees wood has to be permeable particularly for the ascent of sap and the transverse flow of sugar solutions. This section focuses first on longitudinal permeability, though in wood products the transverse permeability plays the more important role. The rise of sap is included as an intriguing phenomenon, and also because it helps explain wood collapse during drying, as well as why some wood becomes impermeable.

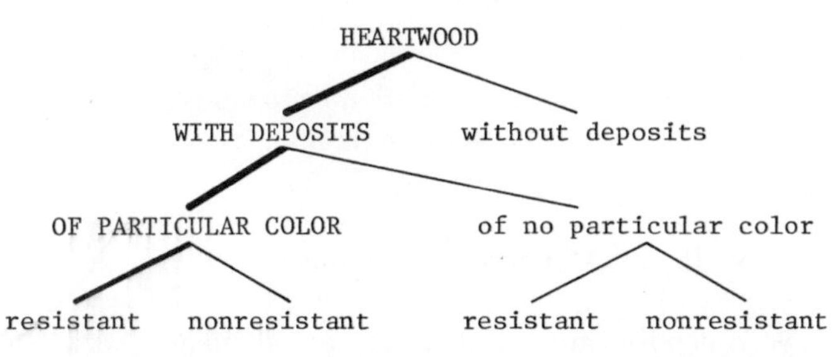

Figure 1.13. Modes of heartwood formation. Capitals and weight of lines indicate frequent types.

Pathway and Ascent of Sap

In hardwood trees sap rises from one vessel member cavity to the cavity of the cell above through perforated plates (Fig. 1.4). In softwood fibers the sap reaches the cavity of the upper overlapping cell through tiny openings, whose structure is illustrated in Fig. 1.14. The walls of the two adjoining cells are pitted, the pits being located exactly opposite each other and separated only by a thin membrane, through whose pores the sap passes. A pair of pits together with the porous membrane is a *pit* in the broad sense. Many pits feature an overhanging rim, which when seen from within the cavity appears to border the pit as a circular band; hence the term *bordered pit.*

The largest openings between vessel members are several thousand times wider than the membrane pores of pits. Since these bottleneck pores control flow, hardwoods possess a much more effective sap-conducting system than softwoods. Here the question arises, how the softwoods can compete and survive among the seemingly better equipped hardwoods. The answer lies in the mechanism of sap rise.

The uphill flow of sap in trees has been a challenge to many minds and was the subject of much controversy, but tree physiologists now agree on the cause. It is in principle the same as when coffee rises in dipped sugar cubes and evaporates at the upper surface of the cube. More comparable to the situation in trees is a capillary tube which pulls the water up (Fig. 1.15). The fluid pressure

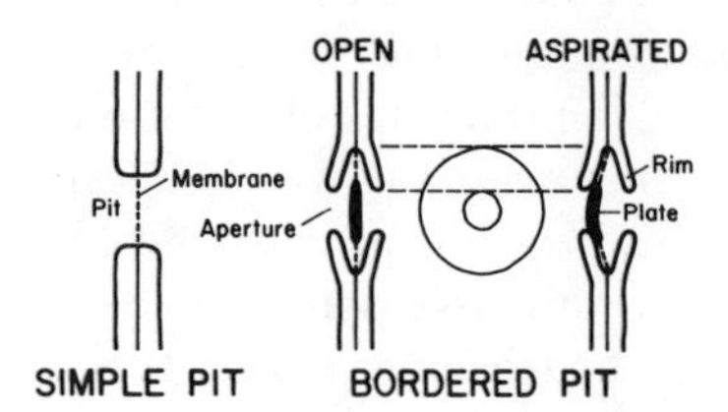

Figure 1.14. Structure and mechanism of openings between cell cavities.

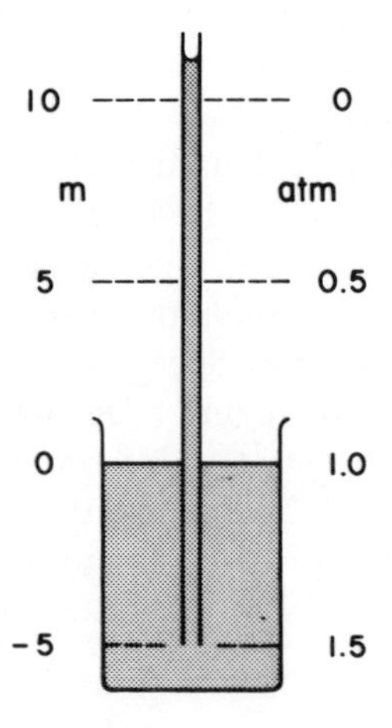

Figure 1.15. Capillary rise and hydraulic pressure in capillary tube.

is one atmosphere (atm) at the level of the container, and 1.5 atm 5 m deep within the fluid. Since outside water and capillary water communicate, the pressure must be the same inside the tube, decreasing with height, reaching zero 10 m above the level of the container to become negative farther up. Ascending water is under a sort of tension stress which can be pictured as reduced cohesion between water molecules.

The tension results from the meniscus at the top of the water column, and depends on the curvature of the meniscus, which in turn is determined by the diameter of the capillary. The capillary rise—11 m in Fig. 1.15—is inversely proportional to the diameter of the tube at the meniscus and does not depend on the diameter of the water column below. Hence the large diameters of **full** hardwood vessels do not limit the rise.

The fact that sap columns end at living cells in leaves rather than in menisci makes no difference with regard to water tension, since living cells create sap tension by means of osmotic forces.

Trees rarely root in water alone, most draw it out of moist soil; they also grow much higher than the 11 m assumed in Fig. 1.15. Therefore sap tension must reach many atmospheres, especially when under dry conditions the leaves temporarily evaporate more water than the roots absorb. Increasing tension or *water stress* causes trees to contract; the trunk circumference fluctuates with the tension during wet and dry periods, as well as in the course of the day, reaching a minimum in the afternoon and a maximum shortly before sunrise. The difference may exceed 1% of the circumference.

Sap tension tends to suck in air from the outside and to create bubbles from dissolved air which disrupt the extremely fragile column, since tension above and below pulls sap away and expands the bubbles. In tracheids of softwoods, menisci in the first-pit membrane block the bubble, confining it to one cell, while vessels of hardwood run completely empty. Hardwoods employ an effective but fragile sap-conducting system, which in species with large pores may cease to function already after one year, to the effect that the youngest annual ring conducts most of the sap.

Pit Aspiration and Tyloses

In addition to the basically safer system, softwoods have an ingenious automatic mechanism for the closing of bordered pits, as illustrated in Fig. 1.14. In the open state, the trampoline-like membrane hovers freely between the two apertures. When one cell cavity empties, sap tension in the other pulls the membrane plate against the rim and obstructs the path to the membrane capillaries (Fig. 1.16). In spruces and several other softwoods *pits aspirate* regularly during heartwood formation and when sapwood lumber dries. The aspiration is permanent; hydrogen bonds hold the plate, just as they hold fibers of paper together, these bonds resisting even high pressure, so that the wood cannot be pressure-treated with preservatives.

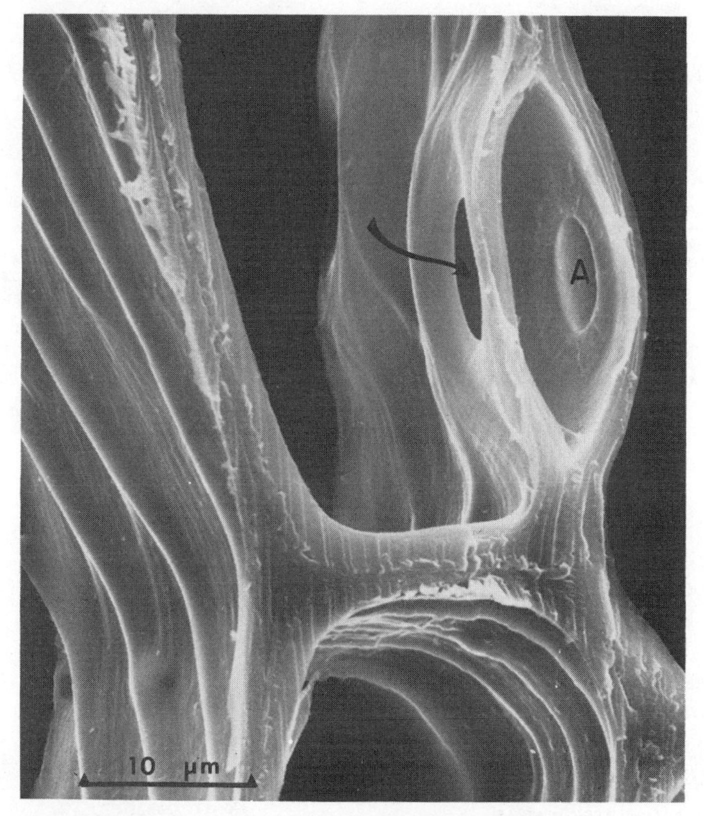

Figure 1.16. Scanning electron micrograph of aspirated pit in Douglas fir. The transparent membrane plate is drawn against the pit aperture (*A*) at right. In the open state water flowed from the cell cavity at left through the pit to the cell cavity at right. (Photograph courtesy of Core, Côté, and Day, 1979.)

Some hardwoods plug vessels as part of heartwood formation. Parenchyma cells around vessel members grow into the vessel, forming froth-like intrusions, which as *tyloses* (singular *tylosis*) fill and plug the whole cavity (Figs. 1.17 and 1.18). White oak species have such ingrowths, whereas red oaks lack them and are therefore not sufficiently tight for beer kegs. On the other hand, tyloses obstruct the flow of preservatives and resist pressure treatment.

Transverse Permeability for Liquids and Gases

Transversely and longitudinally, flowing liquids and gases pass from fiber to fiber through pits of one and the same kind, but transversely many times more pits have to be passed through over a certain distance. The *longitudinal-to-transverse permeability* ratio averages 10:1 for the various wood species. The

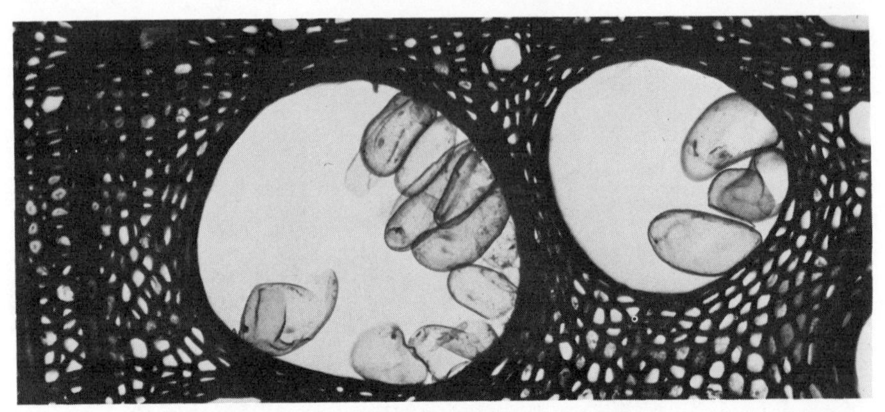

Figure 1.17. Cross section microphotograph of white oak showing tyloses growing into vessels. (Photograph courtesy of U.S. Forest Products Laboratory.)

Figure 1.18. Scanning electron micrograph of white oak cube showing tyloses in four vessels. (Photograph courtesy of I. R. Sachs.)

ratio would actually be much greater if wood did not have additional transverse pathways.

Simple pits at the ends of radially oriented ray cells facilitate the radial flow of sugar solutions from bark into sapwood. For trees the wood must also be permeable tangentially, so that food reserves mobilized early in spring may flow from storage cells to sap-conducting cells. Tangential pathways are also needed during cell growth and for detours of sap around air-filled cells. For these purposes the cells have plenty of simple and bordered pits in the radial cell wall. Even so, wood remains more permeable radially than tangentially, the ratio being in the order of 1.5 : 1.

Some waterproof fabrics *breathe* through pores which are sufficiently large for the passage of individual vapor molecules but not for the molecule conglomerates of liquid water. Wood resembles such fabrics; it is transversely nearly impermeable to liquid water, but not to water vapor and some other gases. For the passage of liquid water, pit openings are barely wide enough, whereas vapor penetrates and diffuses even through the solid cell wall substance, again traveling longitudinally many times faster than perpendicularly where it has to pass more walls.

1.6. CROSS GRAIN

In *cross-grained* wood the fiber direction deviates from a line parallel to the sides of the piece. The deviation reduces strength as explained in Section 4.2, and causes *chipped grain* (Section 11.2), as well as difficulties with glues and finishes which sink into *open* cross grain surfaces.

Disregarding knots, cross grain is either diagonal grain or spiral grain, or a combination of both (Fig. 1.19). *Diagonal grain* results from improper conversion of logs into lumber by sawing at an angle with the bark of the tree, so that in the lumber the layers of annual rings run diagonally through the piece. *Spiral grain* has its origin in the tree: even disregarding trunk taper, the fibers do not run exactly straight up, parallel to the axis of the trunk, but grow in a winding, spiral course about the trunk as if the trunk had been twisted, reminding one of a barber pole (Fig. 1.20). In lumber from spiral-grain trunks the fibers necessarily deviate from a line parallel to the sides of the piece.

Explanation and Degree of Spiral Grain

Some authors see spiral grain as an incidental or accidental effect of tree growth, in spite of the fact that it is **not** related to the rotation of the crown under the influence of prevailing winds, nor to the movement of the earth and the moon. Others, including this author, believe that spiral grain is not incidental, rather that it evolved or was created for the tree's advantage. In perfectly straight-grained trunks each root would supply sap only to those branches that grow out of the trunk exactly above that root-trunk joint. Failure of a main root would shut off sap to those particular branches. Through spiral grain, winding around thin, young trunks faster than around thick, old trunks, each root supplies all branches and each branch receives water from all the roots. In the event that the ground on one side of a tree should dry up, for example under a huge ant-hill, branches on that side might still receive water and survive.

Spiral grain in trees varies a great deal. Practically no trunk is perfectly

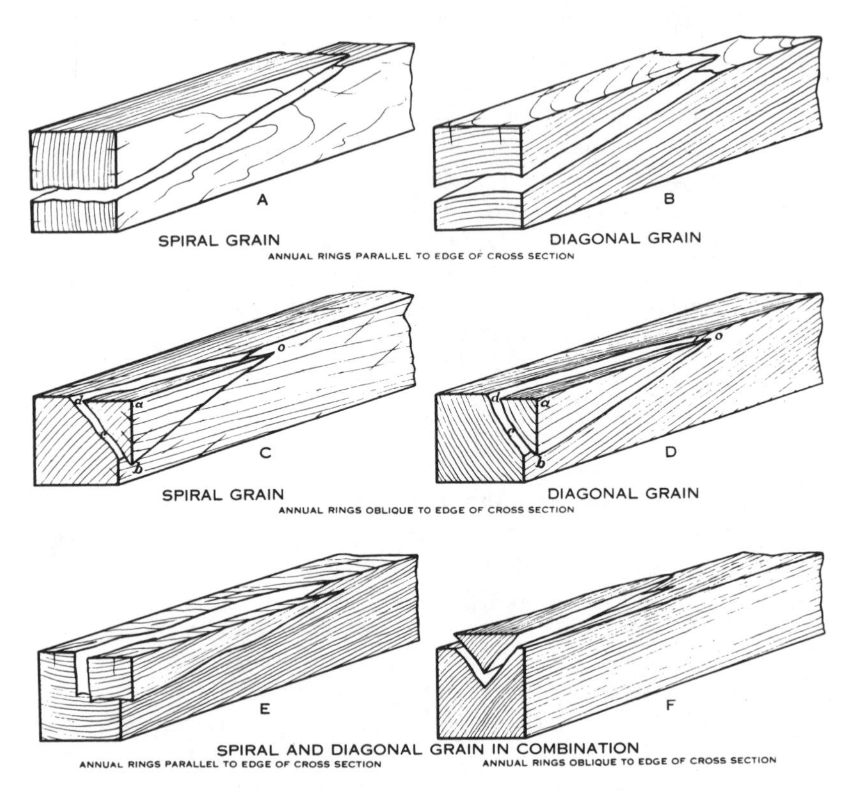

Figure 1.19. Diagrammatic sketches of pieces of wood containing spiral or diagonal grain, and also spiral and diagonal grain in combination. (Photograph courtesy of U.S. Forest Products Laboratory.)

straight-grained. Typically the deviation amounts to a few degrees, but in rare cases the fibers run nearly horizontally, winding at an angle of almost 90° around the stem.

The fibers wind clockwise (left-handed) and counterclockwise right-handed), depending on the tree species, although not every tree follows the rule of its species. In many softwoods the deviation reverses after about 30 years, while numerous tropical species alter the slope in periods of a few years, causing *interlocked grain*, which on radial surfaces looks like ribbons and gives woods such as mahogany unique beauty (Fig. 1.21). Among native trees interlocked grain is common in American elm, sweetgum, and sycamore. Interlocked grain wood has some similarity with plywood: it is fairly strong in all directions, will not split easily, and serves some purposes better than straight-grained wood. But spiral grain of consistent deviation through many years represents a severe defect, in all wood products. Besides having the above-mentioned disadvantages of cross grain, spiral grain causes twisting of drying lumber and of utility poles after installation.

Figure 1.20. Spiral-grained and straight-grained trees. (Photograph courtesy of U.S. Forest Products Laboratory.)

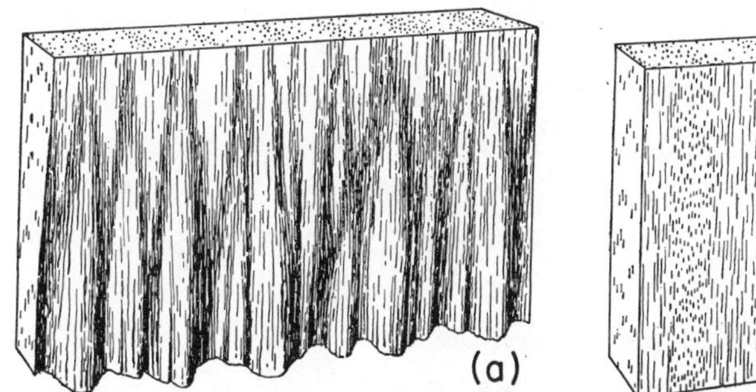

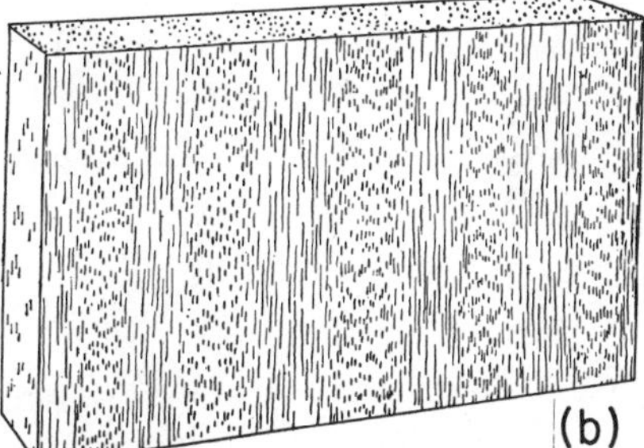

Figure 1.21. Pieces of wood with interlocked grain. (*a*) Split radially with straight edge tool from the top front edge downward; wavy lower edge indicates the depth of interlocking. (*b*) Figure on radially sawed piece. (From Brown and Panshin, *Commercial Timbers of the United States*, 1940.)

Detection of the Fiber Direction

Many pieces of wood *show* the fiber direction while others reveal it in simple tests. In both cases keep two facts in mind: (1) each growing cell lies in the microscopically thin cambial sheath and cannot leave (*run out of*) it; fibers therefore cannot deviate toward pith or bark; (2) within the annual ring layer the fibers may be inclined tangentially to the left or to the right due to spiral grain. Consequently, fiber direction is to be detected in two steps: first find the orientation of the ring layer, then determine whether spiral grain is present within the ring layer (Fig. 1.19).

At clearly tangential and radial sections the orientation of the ring layers is obvious. In nearly tangential sections, short longitudinal distances from ring peak to ring peak as visible in the flat-grained board of Fig. 1.11, speak for layers oriented not quite parallel to the surface. On exactly radial surfaces the ring orientation is identical with the layer orientation, whereas on intermediate sections there are at best approximate clues.

Rays may reveal spiral grain. In exactly tangential sections, the ray spindles naturally follow fiber direction, while on exactly radial surfaces the ray bands appear in full height if the bands (and the fibers) lie in the plane of the section (Fig. 1.5). Bands lower than the known height of the rays speak for fiber deviation. At intermediate sections, shape and orientation of rays tell little about the fiber direction.

Vessels and resin ducts at side-grain surfaces give valuable clues irrespective of ring layers. Long vessel grooves and resin canal lines are telltale signs of the fiber direction. Short grooves and lines indicate that the vessels and ducts *run* either *in* or *out* and that the fibers deviate from the surface plane. Needles forced into end-grain surfaces follow the fibers and may be used to detect the fiber direction.

Splits, drying checks, and pointed tools dragged along the surface in the direction of least resistance all follow the **plane** in which the fibers lie. To obtain the **direction**, open a second plane perpendicular to the first, for example by splitting. In utility poles with drying checks, the second *plane* (actually a round surface) is obviously the ring layer at the pole surface; there the check lines **are** the sought direction.

1.7. KNOTS

Knots add variety to the appearance of wood and cause unique figures, which people appreciate in *knotty paneling* and other rustic products. Wood with tiny knots may be preferred also for resistance against splitting and indentation. Generally, however, knots devalue wood: they are not tolerated in furniture, they render sawing, planing, gluing, and finishing difficult, cause drying checks and warp, deprive lumber of strength, and restrict the use of many pieces. Knots essentially amount to defects, are in fact by far the most common defect in timber, lumber, plywood, and many final wood products.

Knots are bases of limbs imbedded in the tree trunk. As the trunk increases in diameter, trunk wood gradually incorporates the base of the branch and makes it a knot. The fact that each knot is a portion of a branch explains the essential knot characteristics.

Almost all branches originate at the stem's very tip, which later becomes the trunk pith. Only under rare circumstances do new (*adventitious*) branches appear at the surface of thick trunks. Hence practically all knots extend to the pith (Fig. 1.22). Closely related is the fact that branches grow radially or in the radial-longitudinal plane out of the stem. Therefore the radial section of edge-grained lumber exposes branches as elongated *spike knots* (Fig. 1.22), whereas the tangential surface of flat-grained lumber lies perpendicular to horizontal branches and has *round* knots. Knots from steep branches at tangential surfaces, and all knots at intermediate surfaces are between the round and the spike type—that is—oval.

Branches grow in thickness except where imbedded in the trunk, therefore knot diameters increase with the distance from the pith. Knots resemble cones whose tips point to the pith (fig. 1.23). Growth increments of the trunk and of live limbs interlace with each other without forming a line of demarcation: knots of live branches are *intergrown*. At dead limbs the wood formed in the tree trunk makes no further connection with the limb but grows around it, causing an *encased*, loose knot which tends to fall out of lumber.

In branches the fibers run parallel to the axis of the branch as needed for branch strength; hence the direction of knot fibers is entirely different from that of other fibers. In lumber, knots cause excessive cross grain with the associated weakness, drying defects, and difficulties in wood working. The end grain of

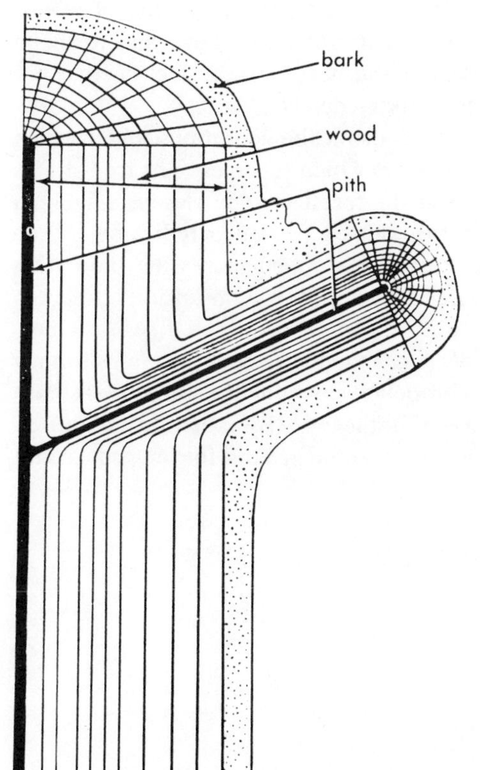

Figure 1.22. Radial section through trunk with intergrown knot. (Adapted from Fames and MacDaniels, *Introduction to Plant Anatomy*, 1925.)

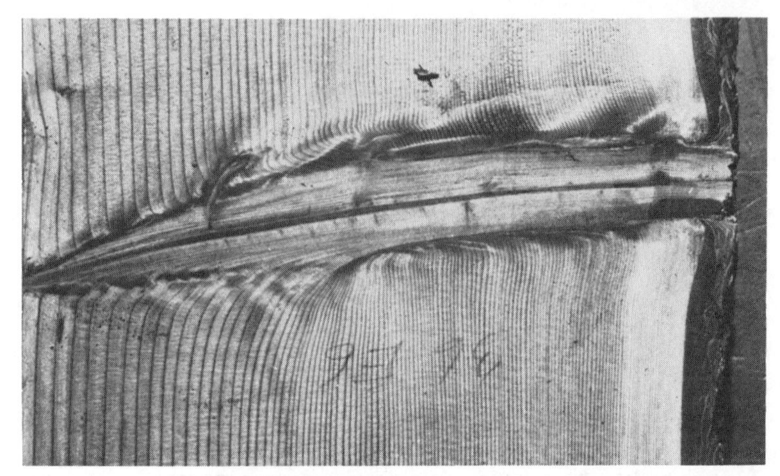

Figure 1.23. Edge-grained Douglas fir board with knot intergrown at left and encased at right. The knot's very tip at left has been cut off together with the pith. (Photograph by B. H. Paul.)

round knots renders gluing and finishing difficult, since the liquid materials sink into the pores, and since exuded resin interferes with finishes.

Branches need more strength than the trunk, not only because of winds, but also their own weight along with the weight of rain, snow, and ice bend them. The knots are correspondingly dense and hard, a fact which further impedes their workability and contributes to drying defects.

Whether lumber is knotty or *clear* depends on the conditions under which the tree grew and on the part of the stem from which it comes. The tree of Fig. 1.23 grew in the open for 25 years. Then the forest canopy closed, neighbors competed for light and nutrients, growth slowed down (as indicated by the narrower rings), the branch died, its diameter stopped increasing, and stem ceased to intergrow with branch. The dead branch stayed on until the tree was felled 90 years later.

Old-growth trees, which for many decades struggled under the crown canopy of a virgin forest, dropped all lower branches early in life; these trees have straight slender trunks free of knots except in the core. By contrast, plantation-grown trees retain branches all the way down to the ground for long times and produce knotty trunks.

2

Diversity, Variability, and Chemical Composition of Wood

Wood differs in the first place from species to species. General species peculiarities are considered here indirectly, in Section 2.2; others appear in later chapters and in the Appendix for wood identification. But wood varies also within each species; just as people who all belong to the same species *Homo sapiens* differ, two trees are never exactly alike. Some particular features of individual trees are genetic like those of the species; others result from the particular environment in which the tree grew. Some within-species variations of both types have already been mentioned in Chapter 1: such wood defects as knots and spiral grain, differences between earlywood and latewood, sapwood and heartwood. This chapter focuses on less obvious variations of *clear* or defect-free wood. Chemically the material differs surprisingly little, but small amounts of certain constituents have great influence on certain properties.

2.1. CHEMICAL COMPOSITION

Knowledge of chemical constituents as provided in this section helps in understanding wood-moisture relations, some gluing and finishing problems, certain differences among the various kinds of wood, and last but not least the reaction wood discussed in Section 2.5.

Wood contains many chemical compounds; few of them originate in the ground, most are formed in the tree. The leaves produce sugars from atmospheric carbon dioxide and soil moisture, using light absorbed by the leaves for energy source, in a process known as *photosynthesis,* meaning *putting together with light*. The inner bark conducts these sugars in aqueous solution to various parts of the tree—mostly to the vast sheath of the cambial layer, whose living cells convert the sugars into the main wood constituents cellulose, lignin, and hemicellulose.

Cellulose and Hemicellulose

Making cellulose from sugars would seem to be easy, since the cellulose molecule represents simply a chain of glucose residues, glucose being a kind of sugar. In reality the process is most difficult, at least for man who so far has been unable to repeat it, although chemists know the configuration of glucose residues in cellulose (Fig. 2.1). The complete molecule of cellulose is a polysaccharide, a compound of many hundreds of sugar residues. Because of the large size of the molecule, cellulose no longer dissolves in water. In cell walls the cellulose chains lie parallel to each other, predominantly in fiber direction for the high strength in that direction needed by the tree.

Cane sugar consists of the two sugar residues, glucose, and fructose. Between cellulose and cane sugar stands hemicellulose, made up of a few hundred sugar residues of various kinds, which vary among the tree species and especially between softwoods and hardwoods. The hemicellulose chains are short and have side chains. Cellulose and hemicellulose make up two thirds to three quarters of dry wood weight, the cellulose portion—around 40%—exceeding that of hemicellulose.

Lignin

Cellulose is the main constituent of all higher plants. Wood differs from other cellulosic tissues in its additional constituent lignin, a compound which makes the tissue *woody* (hard) and imparts strength to carry the tree's heavy crown as well as to resist the high water tension of tall trees. Lignin renders wood difficult to digest for most organisms, giving trees a tremendous advantage over herbaceous (unlignified) plants, which serve as convenient food for thousands of animals and parasitic as well as saprophytic plants. Finally,

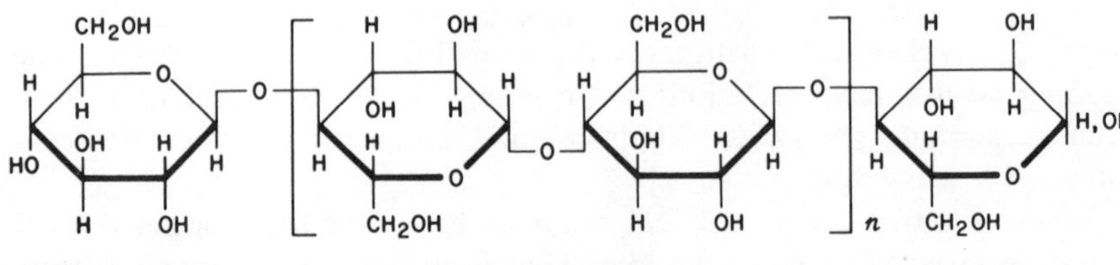

Figure 2.1. Structural formula of cellulose. The number n of glucose residues in the brackets is, for wood cellulose, 500 to 2000.

lignin enables wood to resist the frost of cold winter climates, which withers herbaceous plants to the ground after each growing season.

It is lignification that enables trees to grow higher and older than other plants and to suppress most other vegetations. Lignin makes so much difference and yet the cambial cells synthesize it, surprisingly, from sugars, from the same chemical elements as the carbohydrates cellulose and hemicellulose. Lignin contains more carbon, however, and the elements link up entirely differently in the molecule. Lignin furthermore does not consist of sugar residues and is not a polysaccharide.

Lignin makes wood not only hard but also brittle, breaking under small deformation. Herbaceous stems bend farther than lignified ones; wood does not truly excel on toughness. Softwoods contain more lignin than temperate zones hardwoods, about 30% versus 20%. This high lignin content renders them more brittle and not as well suited for shaping by bending. Hickory as the toughest native wood has relatively little lignin. Tropical hardwoods are still higher in lignin than softwoods and are rarely used for bending.

Distribution of Constituents in the Cell Wall

Cellulose, hemicellulose, and lignin permeate each other. Aggregates of cellulose form a sort of lattice filled with lignin as an incrusting substance. Lignified cell walls resemble reinforced concrete: cellulose corresponds to the steel bars and provides tension strength, while lignin in the role of concrete makes the framework hard, stiff, and resistant against compression. Hemicellulose seems to supplement the function of cellulose.

The three constituents are distributed unevenly in the wall. Lignin is plentiful near the joints of cells, functioning not only as an incrusting but also a cementing substance. When converting wood into pulp for good paper, the lignin has to be dissolved to separate the fibers, and must then be leached out of the cell walls to render the fibers flexibly soft.

Extraneous Wood Constituents

Extraneous constituents are by definition not an essential part of wood and do not occur in all wood; most species contain a few percent at the highest. The compounds are believed to be made also from sugars, some possibly in the leaves, others in the dying parenchyma cells of the sapwood-heartwood transition line. Extraneous constituents occur in cell walls, in cell cavities, and in intercellular spaces such as the resin canals. Most, the *extractives,* can be removed with neutral organic solvents, some even with water.

The heartwood deposits mentioned in Section 1.4 are—besides resin—the most significant extraneous contituents and give the heartwood of many tree species color and durability. The poisonous nature of some tropical wood species results from poisonous extraneous constituents. Certain stains, finishes, and glues may react with the extraneous substances and change the wood color. The tannins in oak and other species combine with iron to form a sort of ink that causes unsightly black stains, but only in moist wood, because the reactants must be dissolved.

The food reserves mentioned earlier are a special group of extraneous constituents; they originate from sugars, which living parenchyma cells in sapwood convert to starch for storage. Starch does not dissolve in cold water and can be reconverted to glucose when needed—characteristics ideal for storage. Starch is indeed plentiful in plant tissues whose main purpose is to store food, such as fruits and the various tubers, like potatoes.

Resin

Many softwood tree species seal wounds and suffocate invading organisms by means of resin, a gummy yellowish substance which is produced in sapwood parenchyma cells. Most of the resinous species store and transmit the substance in *resin canals* or *resin ducts,* a sort of intercellular, parenchyma-cell-lined passageways. The cross sections of Figs. 1.1a and 1.3A include longitudinal resin canals, other resin canals run radially within rays. The longitudinal canals are in several species sufficiently large to be visible to the unaided eye, appearing at smooth cross sections as small dots, and at planed lateral sections as thin streaks. Resin canals are never as numerous as vessels and hardly contribute to wood's figure; even in large numbers they would not affect wood permeability because they are plugged by viscous resin. Hardwoods lack canals of this type, but several tropical species feature corresponding gum canals.

Resin contains turpentine and other solvents which dry out and give the species a characteristic resinous, in pines a *piny* smell, while the residue slowly hardens. At high temperatures not only the solvents but also some resin solids evaporate. For some uses of wood, such as windows, relatively high resin contents are desirable since to some extent the resin seals out water, contributing in this way to wood durability; but the resin may interfere with finishing (Section 14.3).

2.2. DENSITY AND POROSITY

Density is the ratio of the mass of an object to its volume. When applying the term to wood, we consider the cell cavities as part of the volume and express density in pounds per cubic foot, gram per cubic centimeter, or kilograms per cubic meter. Density in units of gram per cubic centimeter is numerically identical with *specific gravity,* the ratio of a particular density to the density of water (which happens to be 1.00 g/cm^3).

Density affects the cost of transport and determines the force required to lift wood; more importantly it serves as a measure of other properties, representing the simplest and best indicator of wood quality in general. Wood technologists memorize the specific gravities of the various wood species more accurately than any other characteristic. When confronted with an unknown wood, simply *weigh* the piece in the hand to estimate strength, workability, drying characteristics, and most other properties.

Effect of Moisture on Density, Specific Gravity

The moisture present in practically each piece of wood influences density in two ways: through the mass and by swelling, each by different percentages so that density changes with the moisture content as illustrated in Fig. 2.2. The relationship appears to be curvilinear because the moisture scale is logarithmic; on a linear scale the lines would be straight with a shrinkage-related bend near 30% moisture content (Chapter 3). Unless stated otherwise, I shall give densities at 12% (D_{12}), the average moisture content of wood in use.

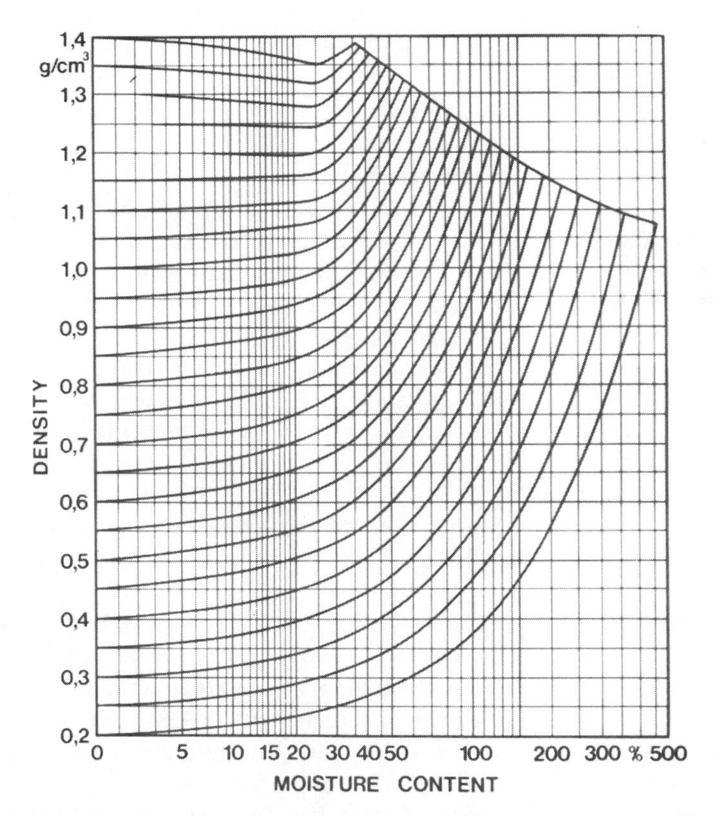

Figure 2.2. Relation of density and moisture content. Example: wood that has a density of 0.5 g/cm^3 at 0% weighs 0.6 g/cm^3 at 35% moisture content. (By F. Kollmann; photograph courtesy of D. Noack.)

In the wood technology literature (but not in this book), *density* has a somewhat different meaning: there it is defined as the ratio of *ovendry weight to volume at a specific moisture content,* the moisture content being either *green*—the state in the living tree (far above 30%; see Section 3.4)—or 12%, or any other. Wood technologists prefer the corresponding term *specific gravity* (G), which in this sense is the ratio of the *ovendry weight of a sample to the weight of a volume of water equal to the volume of the sample, at the specified moisture content.* I shall use *specific gravity* in this sense, mostly based on volume at 12% moisture content.

For the conversion of specific gravity G into density D (g/cm^3) apply these equations:

$$D_{12} = 1.12\, G_{12} \qquad D_{30} \approx 1.3\, G_g$$

$$G_{12} \approx \frac{G_g}{(1 - 0.17\, G_g)} \qquad G_g \approx \frac{G_{12}}{(1 + 0.17\, G_{12})}$$

D_{30} stands for the density at 30% moisture content, while G_{12} and G_g are the specific gravities on basis of volume at 12% moisture content and in the green state, respectively. More accurate values of D_{30} and G have to be calculated by using shrinkage data for the species (Tables 4.1 and 4.2).

Cell Wall Specific Gravity and Wood Porosity

Dry wood consists of cell walls and air-filled cell cavities. Since air has practically no mass, the weight of dry wood equals the cell wall weight, which depends on the proportion or volume of the wall and on the wall's specific gravity. Fortunately cell-wall specific gravity is practically same for the various kinds of wood cells, for earlywood and latewood, in sapwood and heartwood, and even among the various tree species: it is always near 1.5 on basis of ovendry volume. This figure allows us to calculate how much volume the wood substance occupies in wood of known specific gravity G: the ratio of *cell-wall volume to wood volume* equals the ratio G:1.5. In Douglas fir for example, with an average specific gravity near 0.5, cell walls comprise 0.5:1.5 or one third of wood volume. Two thirds of Douglas fir is cell cavities, open to wood preservatives, glue, paint, and water.

Conversely, the wall proportion visible in microscopic wood cross sections *shows* the density. In the cross section of northern white pine in Fig. 1.3A, cell walls occupy at least one fifth of the area, corresponding to the specific gravity one fifth of 1.5, or 0.3. The species averages 0.35. Since specific gravity is an indicator of many wood properties, the microscopic view indirectly tells much about the wood even if the species cannot be recognized.

The specific gravity G_{12} of the lightest wood species, balsa, averages 0.17, corresponding to almost 90% cell cavity space in dry pieces. Dry balsa is as buoyant as cork and is used for rafts in South America. The species' name comes from this fact, the Spanish substantive *balsa* meaning *float, raft.* Lignumvitae, another species from Latin America and known as wood for bowling balls, ranks as the heaviest commercial wood species, with a specific gravity G_{12} of 1.09; only one third of its green volume is available for conducting sap.

Average specific gravities G_{12} of commercial native species range from 0.31 of northern white cedar to 0.88 of live oak. The values vary within annual rings, in various parts of the tree, from tree to tree, from one forest site to another, and may differ in various geographic regions. In a shipment of Douglas fir for example the lightest stud may measure 0.45 and the heaviest 0.55. Specific gravities of important wood species appear later in Tables 4.1 and 4.2 together with mechanical properties.

2.3. DOES RING WIDTH INDICATE QUALITY?

Density indicates wood properties only if the moisture content is known. What is more, the weights of logs, stacked lumber, or large structural members cannot be estimated by simply lifting them. In such cases the ring width at the end face and at exactly radial sections may give a clue as to whether the piece is below or above average density.

Laymen indiscriminately associate wide rings with low density and weakness, because aspen, yellow poplar, the cedars, and other fast-growing species have wood of low density; so they conclude that fast-grown wood is relatively light. In certain species wood with wide rings is indeed relatively light. Southern yellow pine *structural* lumber, for example, qualifies for the strong *dense* grade if it has at least six rings per inch. However, even within these species the rule has limitations.

Several other softwoods, especially the sunlight-demanding species, respond to the rate of growth in a similar way (Fig. 2.3). Specific gravity of the commercially very important Douglas fir peaks at about 30 rings per inch, but

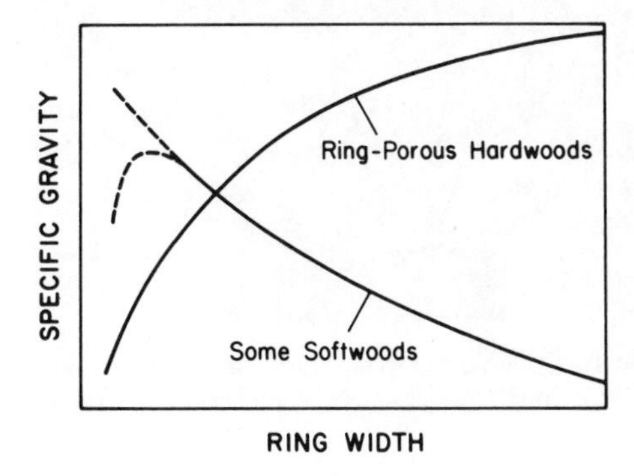

Figure 2.3. Relation of specific gravity and ring width.

the species takes many exceptions to this rule. In other softwoods the relationship between ring width and specific gravity is even more complex.

The specific gravity of diffuse-porous hardwoods shows no relation to growth rate, whereas that of the ring-porous group increases with increasing ring width (Fig. 2.3). In the latter group, porous earlywood bands comprise almost the whole narrow ring, whereas in fast-grown trees the earlywood band is hardly wider than in slow-grown trees so that the growth ring consists mainly of dense latewood (Fig. 2.4). In reality, the proportion of earlywood and latewood also determines how ring width influences the specific gravity of softwoods. Southern yellow pine with wide rings for example tends to have high proportions of earlywood and to be correspondingly light for this reason. Since the diffuse-porous hardwoods show no difference within the annual ring, it could be expected that their specific gravity does not correlate with the growth rate.

2.4. VARIATION OF DENSITY WITHIN TREES

Organic tissue generally changes with the age of the growing organism. It is of interest to know whether similar differences exist also in trees, particularly whether wood properties depend on the tree age at which the wood grew. The

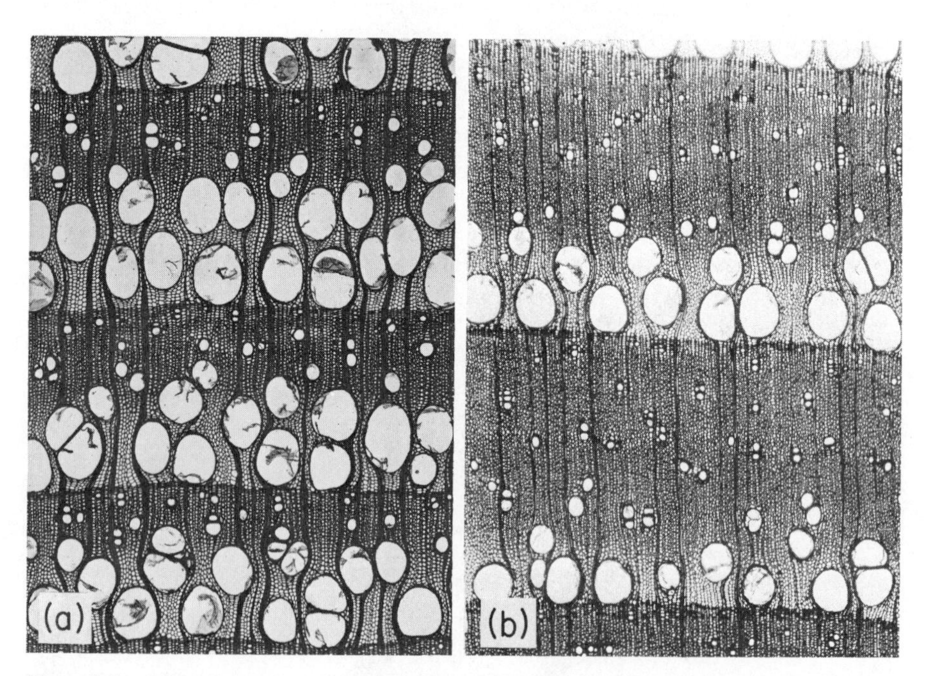

Figure 2.4. Effect of ring width on specific gravity of ring-porous wood, shown in cross sections of white ash. (*a*) Narrow-ringed light wood. (*b*) Wide-ringed dense wood. (Photograph courtesy of U.S. Forest Products Laboratory.)

question has significance, since curvature and width of annual rings *show* the approximate age of each piece. Whether the height in the tree influences wood properties is less interesting, because in lumber and other wood products the height is no longer visible. Actually, properties of clear wood **do** differ within the tree. Some of the variations follow regular patterns.

In many trees the wood of the first 5 to 20 annual rings around the pith is *juvenile,* that is to say, relatively light, having short fibers, and containing less cellulose than the *adult* wood of later annual rings. The properties of juvenile wood change progressively with increasing distance from the pith until they reach the fairly uniform level of adult wood. Tissue of this kind always seems to grow in the tree crown, but the tree usually progresses from *juvenile* to *adult* wood production before its branches dry up. Most tree species and nearly all trees within these species have juvenile wood, no matter how fast they grow. In some species the juvenile wood may lack one or two of the three characteristics—low density, short fibers, little cellulose—but all three characteristics have the same effect: they cause weakness. In addition, coniferous juvenile wood tends to shrink excessively in fiber direction and contributes to the warp of lumber during drying.

Adult wood still varies with the distance from the pith, but less than juvenile wood. The changes do not follow a simple pattern, but two major groups of species show some regularity. Since annual growth tends to diminish as trees grow old, wood formed late in the life of ring-porous hardwoods is relatively light. Conversely, in most sunlight-demanding conifers, density may peak in the growth increments of mature trees (compare Fig. 2.3).

The width of each annual ring sheath also depends on the height of the tree. The rings are narrow near the base of the trunk and become wider toward the crown, reaching a maximum in the vicinity of the crown base, narrowing again towards the top. This growth pattern causes the stem to be cone-shaped (tapered) in the crown area and nearly cylindrical in the trunk below the crown. Because of the influence of height, new wood in ring-porous hardwoods is relatively dense at the base of the crown, whereas in sunlight-demanding softwoods it is frequently relatively light.

2.5. REACTION WOOD

Many of the undesirable variations within trees result from certain environmental conditions and changes in the environment. Very drastic changes of a certain kind lead to formation of reaction wood tissue, which is a major cause for warp in lumber.

Occurrence and Mechanism

Reaction wood is *wood that reorients the tree* into more favorable positions. We know that house plants, after being turned away from the light, bend back within a few hours, days, or weeks. Responses of this kind by trees are hard to imagine because bending thick trunks requires tremendous forces, but trees do generate them, although slowly through many years. Old Russians must have observed this, since according to one of their numerous proverbs *trees always grow upward no matter how one bends them.*

Figure 2.5 illustrates a common situation in forests. After the fall of one of the dominating trees, the suppressed conifer bows into the opening by means of excessive, one-sided growth of reaction wood, which generates longitudinal stress and forces the stem to bend over. Opposite to the reaction wood flank, much smaller amounts of normal wood grow in eccentric annual rings. The tree grows reaction wood all along the right flank of the stem until the crown attains optimal exposure to light. How many years this reorientation takes depends on the diameter of the trunk. Saplings respond noticeably within a few days and complete the bending within one year, while trunks over 10 in. in diameter need several decades to attain the favorable position, and old trees never do make it completely.

The spruce cross section of Fig. 2.6 bears evidence of a forest disaster. The tree grew straight and vertically until it was uprooted in a rainstorm in its 23rd year; a gust blew it down without breaking the trunk, but tore the root plate up from the ground, all except one edge. The tree tried to bend up and grew reaction wood at the lower flank (bottom in Fig. 2.6), while growth of normal wood on the upper flank nearly ceased. When cut down seven years after the storm, the upper part of the old stem had regained the vertical position, but was still low, since near its root plate the stem was still horizontal. The more fortunate neighbors had grown into the opening and had shaded the victim out; hence its diminishing growth during the last years.

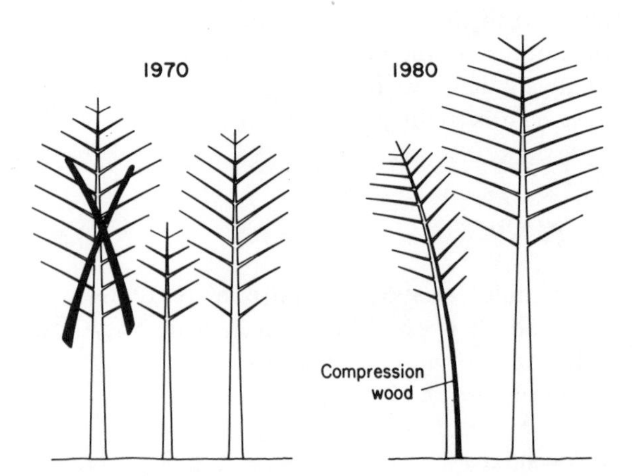

Figure 2.5. Formation of reaction wood. After removal of the big tree at left, the suppressed conifer bends to the light by growing compression wood.

Figure 2.6. Cross section of white spruce trunk with distinct compression wood.

Most stems with bends contain reaction wood, but this abnormal tissue also occurs in straight stems that in a repeatedly changing environment bend for a few years to one side, then for another period to the other side, and so on, remaining fairly straight all the time. Such behavior is typical for aspen and other species of the genus *Populus*; their reaction-wood flanks tend to alternate, the tissue occurring in little patches, rather than in crescents of eccentric growth rings.

Reaction wood appears as *compression wood* in coniferous and as *tension wood* in broad-leaved trees. The conifers reorient themselves through longitudinal compressive stress as described above, with wood at the outer (convex) flank **pushing** the tree over or up. Broad-leaved trees **pull** themselves over or up by means of longitudinal tension at the inner (concave) flank. A simple model illustrates how tension wood works: press stretched masking tape on a thin plastic ruler—the ruler bows. To simulate compression wood, bond a soft, lengthwise compressed rubber bar to the ruler. The stress in tension wood is due to fibers attempting to contract longitudinally during maturation, while older wood restrains this contraction. Correspondingly, it can be assumed that compression wood attempts to expand in fiber direction when lignin forces cell wall components apart; compression wood indeed contains much more lignin than normal wood.

Reaction Wood Properties

In Fig. 2.6 lignin-rich compression wood appears darker than normal wood; in reality it looks reddish and is therefore called *Rotholz* in German, literally meaning *redwood.* Compression wood features relatively large proportions of latewood and frequently lacks the demarcation between earlywood and latewood. Tension wood appears less conspicuous; being rich in cellulose, the constituent that provides tensile strength, it is generally relatively light in color and has a sort of sheen when observed at certain angles. Tension wood fibers hold together so tenaciously that sawed surfaces often look wooly or fuzzy (Section 11.3).

Reaction wood shrinks and swells excessively lengthwise—in softwoods up to 10 times more than normal wood. This causes lumber, in which the tissue as usual occurs in one flank only, to warp and sometimes even check perpendicular to fiber direction. Aspen lumber always seems to contain tension wood; studs of this species warp as a rule. Long two-by-fours are never perfectly straight in any species, the warp frequently being caused by reaction wood of various intensities. Due to tenacious fibers, excessive shrinkage and the resulting internal stress, saws pinch when ripping lumber containing the abnormal tissue. Reaction wood exceeds normal wood in density and in strength, as might be expected from its function in the tree, but curiously enough it is not as strong as normal wood of the same density, and may even be weaker in certain kinds of strength.

3

Wood-Moisture Relations

Moisture impairs wood's dimensional stability, strength, and durability; it affects sawing, planing, finishing, and all other kinds of wood processing. The integrity of glue bonds depends on wood moisture at the time of the gluing operation and thereafter. Moisture likewise influences paint durability. It is therefore vital to gain control of wood moisture, learn how to measure it, learn its effects and how to prevent moisture-related difficulties.

3.1. HUMIDITY AND VAPOR PRESSURE

Wood moisture depends, quite simply, on atmospheric humidity. Vapor in the air influences the drying rate, determining how far wood dries and how much moisture wood contains in buildings. We need some knowledge of humidity to understand wood-moisture relationships.

In the air invisible vapor molecules scatter or diffuse about in an irregular fashion in all directions. Molecules that incidentally come close attract each other, but their kinetic energy prevents them from cohering. However, when the molecules are crowded and collide frequently enough with each other, their kinetic energy is insufficient to overcome mutual attraction, and so the vapor condenses to form water droplets which appear as fog. Air has only limited space for vapor. We perceive the degree to which vapor fills the space as humidity. *Relative humidity* (RH) is *the percentage of vapor the air is holding, in relation to the amount it is able to hold at a given temperature.*

Molecules of nitrogen and oxygen have no influence on the attraction between vapor molecules. The greatest amount of vapor possible in a given volume is the same whether air is present or absent. Air does not *hold* moisture, the space does. The designation *saturated air* is misleading; *saturated vapor* describes things better.

At higher temperatures, vapor molecules—like all gas molecules—diffuse and scatter more vigorously. In fact, temperature is a measure of molecular movement. Warmer molecules have more kinetic energy to escape each other's attraction and can be closer without condensing. The amount of vapor that air can contain increases with the temperature of the air. Hence in a room or space containing a certain amount of vapor, RH decreases as temperature rises and vice versa; RH therefore rises during the night and drops after sunrise. The *dew point* is *the temperature at which vapor begins to condense in cooled air.*

The air in heated buildings is dry because the cold air outside, which infiltrates the building, contains little moisture, no matter how high the RH outside may be. The furnace draws air as it consumes oxygen and thrusts gases out through the chimney; equal amounts of air enter through unsealed windows and through open joints in the structure. This dry infiltrating air picks up whatever moisture evaporates in bathrooms, in kitchens, in lungs, from the skin of occupants and from plants, carrying it away through the furnace.

Atmospheric pressure [about one atmosphere, 14.7 $lb/in.^2$ (psi)] is a result of air molecules hitting objects in great numbers each second; it is the sum of all these minute impacts of nitrogen, oxygen, and other air molecules, among them vapor molecules that exert *vapor pressure.* The latter naturally depends on the number of vapor molecules in the air and is proportional to RH. In fact, physicists define *RH* as *the ratio of the actual vapor pressure to the pressure of saturated vapor.*

Vapor pressure (p_s) for saturated vapor at various temperatures are listed in handbooks of physics, chemistry, and engineering. With p_s we can calculate the vapor pressure (p) in any atmosphere of known RH: $p = p_s \cdot \mathrm{RH}/100$. Vapor pressure indicates the direction of moisture movement, since vapor diffuses toward the area of lower pressure. Actually the molecules diffuse in all directions, but a majority scatters to the area where less molecules are, where vapor pressure is lower. In heated homes, evaporation of water raises vapor pressure to a high level compared with the vapor pressure outside, so that molecules migrate through gaps along windows and doors, through porous walls and open doors to the outside, in this way contributing to the dryness of air inside.

The vapor above water in a closed container is saturated; its pressure can be attributed to the liquid. Water has the same vapor pressure as saturated vapor at the water temperature; it has this same pressure also in (wet) wood. Hence the vapor pressure of wet wood equals known vapor pressures p_s of the air.

One may therefore apply the vapor pressure concept to drying, to the migration of moisture in outside walls and doors, and to the distribution of moisture in buildings.

3.2. EQUILIBRIUM MOISTURE CONTENT

Strictly speaking, the wood in our surroundings is always moist. Moisture content varies with conditions, causing the material to swell and shrink; sometimes the moisture reaches the degree of wood rot. In this section I shall explain why wood contains moisture, how much, and what causes the fluctuations.

Nature of the Moisture Equilibrium

Trees synthesize sugars and wood constituents in aqueous solutions of protoplasma in leaf, cambium, and wood parenchyma cells. Water carries the synthesized compounds in solution to other parts of the tree, playing a major role in all phases of wood growth. The constituents have an innate affinity to water and do hold water.

Wood seems to attract water out of the surrounding atmosphere. Some diffusing vapor molecules that hit the wood surface get caught in the attracting range of wood constituents. Wood is a *hygroscopic* material (from Greek *hygro*, moist, and *skopein*, looking for).

According to the well established *kinetic theory*, molecules of all matter are in continuous motion, even in solids where they vibrate in certain positions, swinging back and forth between their neighbors. The wood moisture molecules considered here behave like molecules of solids; some—vibrating more than others—escape from the attracting forces of the wood constituent, shoot off, lose their kinetic energy in colliding with another wood molecule, and get trapped again to vibrate at the new site. But a few reach the surrounding air; some moisture escapes from wood all the time.

Whether arrests or escapes of vapor prevail depends on the vapor pressures of wood and of the air. In dry wood surrounded by humid air, arrests exceed escapes and the wood gains moisture, whereas in moist wood exposed to dry air the escapes dominate and the wood dries. Between these is a state in which escapes equal arrests: for each RH there exists an equilibrium moisture content (EMC), as illustrated in Fig. 3.1.

In the cell wall, moisture occupies certain sites that determine the wall's moisture holding capacity. In *fiber saturated* wood, water has entered all the available sites. The *fiber saturation point* (FSP) is near the moisture content (MC) of 30%, the EMC for the highest possible humidity.

Variability of the EMC

The EMC is the same from piece to piece, even from species to species. Samples in a 65% RH room always approach the equilibrium of 12% MC when being conditioned for standard tests. The EMC is the same because wood species differ little in chemical composition, and because the different wood

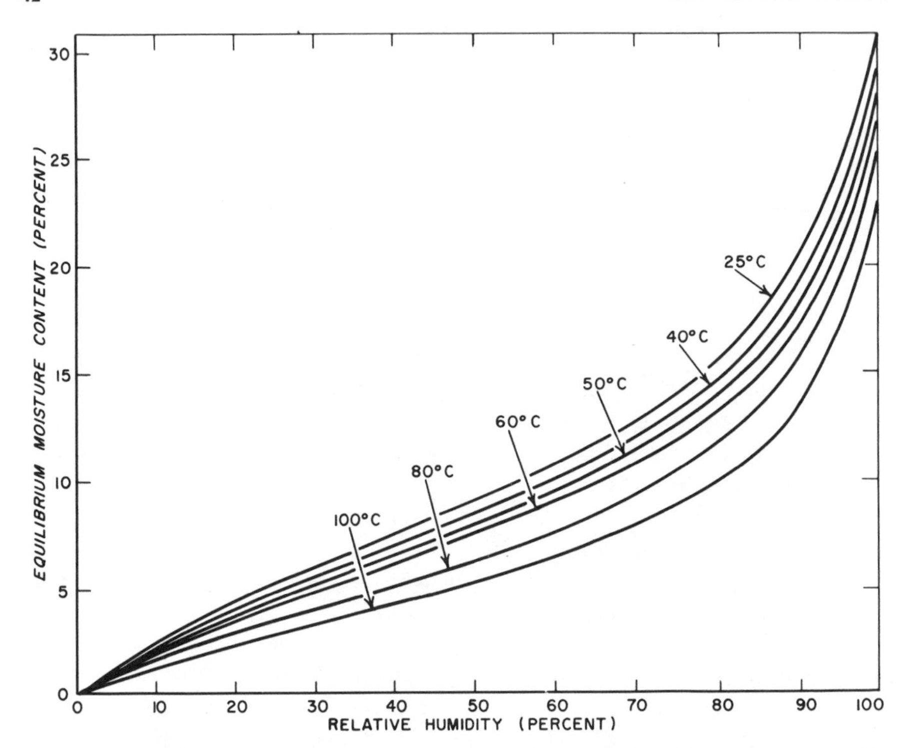

Figure 3.1. Relation of equilibrium moisture content and humidity for various temperatures. (By U.S. Forest Products Laboratory; from Koch, 1972.)

constituents are very similar in hygroscopicity. High percentages of lignin for example hardly alter the EMC.

Only woods that contain relatively large amounts of extraneous constituents in cell walls deviate from the EMCs in Fig. 3.1. Many constituents of this kind occupy some moisture sites and *bulk* the wood substance, reducing the EMC and particularly the FSP. Therefore the FSP of heartwood is lower than that of sapwood in many wood species, the difference being a few percent of moisture content. The extraneous constituents reduce the moisture contents for low humidities less than they do the fiber saturation point; at 0% relative humidity the effect must be zero. Redwood, western red cedar, and teak contain unusually large amounts of these constituents; their heartwood EMC for 65% RH is near 9%, not 12%.

To some extent the EMC depends on whether the wood reached that level from above or from below, being higher if attained during drying (*desorption*) than after moisture gain (*adsorption*). The equilibrium has a *hysteresis* (from Greek, a coming short, deficiency) as a result of inertness or sluggishness of the material. (If a container of sand for example tilts to one side, the sand shifts; when restored to original upright position the sand shifts back but not completely, some remains shifted.) For the moisture contents reached in very dry air the hysteresis is insignificant (zero at 0% RH), but in humid conditions the difference between EMC of adsorption and desorption reaches several percent MC.

Wood's EMC varies little with temperature; it is almost the same for example at 40°C as at 25°C (Fig. 3.1), though at the same RH hot air *holds* much more vapor than cold air.

The wood-base panel products—plywood, particleboard, and fiberboard—resemble wood in hygroscopicity; the amounts of binder used are too small to alter the EMC significantly. However, very high manufacturing temperatures cause chemical changes which show in darker colors and involve lower equilibriums. These extreme chemical changes may reduce the EMC at 65% relative humidity, for example, from 12% down to 8%.

Wood is by no means unique in hygroscopicity. Other organic materials—including hay, corn, peas, honey, meat—are also generated in aqueous solutions of protoplasma and have a similar affinity to water; their EMCs are in the same order of magnitude as those of wood. Fats and oils as water-repellent substances are the exceptions.

3.3. DETERMINATION OF MOISTURE CONTENT

Wood technologists express the amount of water in wood as a percentage of the weight of dry wood (W_d) and define the moisture content (MC) as

$$\mathrm{MC} = \frac{(W_m - W_d)}{W_d} \cdot 100\ [\%]$$

W_m stands for *moist weight.* Wood that contains a weight of water equal to itself has a moisture content of 100%. More water than wood means a percentage greater than 100. Wood may indeed have several hundred percent MC.

Ovendrying in Principle

The simplest accurate way for the determination of MC is ovendrying. Its procedure follows from the given definition of MC: obtain the initial moist weight (W_m), place the sample in a hot oven, weigh repeatedly to the constant weight (W_d), and calculate the MC. Many manufacturers and consumers of wood products, suspecting their material is too wet or too dry, are not aware of the ease of this method.

The *ASTM* standard *D 2016* (see the *ASTM* book mentioned in the Bibliography), describes the ovendrying method in every detail. The oven temperature should be regulated between 213.8° and 221°F (101° and 105°C), a requirement that laboratory ovens meet. Besides raising wood's vapor pressure and so accelerating the drying many times, the heat reduces relative humidity.

Assume the vented oven is located in a room of 70°F (21°C) and 40% RH. Heating to 217.4°F (103°C) lowers the humidity in the oven to 0.83%. Because of this trace of humidity the sample will retain a minute amount of moisture; however, the standard D 2016 defines the state reached in air of 217.4°F (103°C) as *ovendry* (0% MC), regardless of the relative humidity outside the vented oven.

Small temperature deviations cause only minor errors, so that for many purposes drying in household ovens or even on heat radiators is sufficiently accurate. Take the example of a 40% RH room; an oven temperature of 176°F (80°C) lowers the RH not to 0.83% but to 2%, with the effect that wood retains 0.15% more moisture (again based on ovendry weight) than at 217.4°F (103°C). Above 221°F (105°C) some wood constituents decompose and escape as fragments into the air together with the moisture; weight of the sample then does not reach constancy, and the determined MC becomes inaccurately high. Short exposures to a few degrees of excessive heat causes only minor errors (compare Section 8.2). Strictly speaking, even at 213.8°F (101°C) some wood constituents decompose or oxidize, with the effect of slowly decreasing sample weight. Therefore avoid drying for periods longer than necessary to achieve an approximate constant weight.

Volatile extraneous constituents evaporate along with the moisture especially at high drying temperature, again leading to excessive moisture contents. Consider this when dealing with softwoods like southern yellow pine that contain turpentine as a resin solvent.

Samples for Ovendrying

The test specimen may have any size, ranging from a tiny sliver to a whole board, but it must suit the available scales and the drying oven. In most cases 2 to 20 $in.^3$ samples serve the purpose. Small specimens call for special care to prevent their drying before weigh-in. On the other hand, drying large samples may take too long, inasmuch as even a 2 $in.^3$ piece that is already air-dry requires about 24 hours. Repeated weighing during drying becomes unnecessary after one learns how long particular samples dry under given conditions.

To determine the average MC of a shipment of lumber, pile several randomly chosen boards above each other, bore through them with an *increment borer,* and weigh the borer core either as a whole or as individual pieces. Increment borers—designed to help foresters evaluate growth increments in trunks of standing trees—are well suited for this purpose, as the bored-out core does not begin drying inside the cylindrical borer shaft until pulled out with the *extractor.*

Since surfaces and ends of lumber dry faster than centers, the moisture content varies within the piece and the sample should be taken as a full cross section far from the end. To determine the moisture gradient across thick lumber, slice a short section into several layers parallel to the lumber surface.

Other Methods

The *ASTM* standard *D 2016* includes electrical moisture measurement and several other methods. The electrical meters test an electrical property of wood that depends on moisture content; being calibrated in percentages of moisture they are very convenient. Unlike ovendrying of samples removed from lumber, this method measures *nondestructively*, leaving the wood undamaged. On the other hand, meters cost money, from $85 up, and do not give very accurate readings.

Most electrical moisture meters rely on the fact that moisture lowers the electrical resistance of wood. Dry wood is an electrical insulator, whereas wet wood—or rather the water in wet wood—conducts electricity, although not as well as metals do. In the moisture range of interior woodwork, where loss of 1% of moisture causes the resistance to double, the resistance meters deviate from true value by at most a few MC percent. Below 6% MC however the resistance becomes too high, and above 25% it changes too little for accurate measurement with this type of meter. The manufacturers provide correction tables for various wood species, lumber thickness, and temperature.

The moisture content of wood containing volatiles, such as the wood preservative creosote and the turpentine of resin, is determined accurately by distilling the moisture from the fractured specimen along with toluene or another liquid that is immiscible with water, catching the water in a trap, measuring its volume, and calculating the MC. Laboratory glassware required for this test is inexpensive.

3.4. FREE WATER AND MOISTURE CONTENT OF GREEN WOOD

Section 3.2 concerned *hygroscopic* or *bound moisture* in cell walls, as opposed to *free water* in cell cavities. The two differ essentially: bound moisture affects almost all properties of wood, while free water only adds weight and prolongs drying. In this regard wet wood resembles a water-filled paper cup: water in the cup exerts no influence on cup strength, but moisture that penetrates the paper weakens the cup.

In dry wood exposed to humid air the cell walls adsorb moisture while the cell cavities remain empty (the small amount of vapor in cell cavities can be ignored compared with the mass of bound moisture). The cavities remain empty even at 100% RH where the EMC reaches the maximum at fiber saturation (Fig. 3.1). To really fill the cavities, liquid water must wet the wood surface and soak in through the exposed cavities and pit openings. In wetted dry wood the cell walls are the first to adsorb the penetrating liquid, and they stop adsorbing only when they reach fiber saturation.

Conversely, in the drying process, free water evaporates first before the cell walls begin to dry. Of course, this applies to single cells and small groups of cells only. Cell walls at the surface of wet lumber **do** lose bound moisture in dry air; the surface layers dry below fiber saturation before all free water has migrated from center to surface.

The young wood cells of living trees are full of water, otherwise their leaves could not pull up sap. Dead pieces of wood reach this *water-soaked* state when boiled or when water is pressed into evacuated material, or when soaked under water for many years. How much water the soaked wood contains depends on cell cavity space, which in turn is a function of specific gravity. Theoretically a wood lacking cell cavities (wood of specific gravity 1.5, equal to the specific gravity of cell wall substance) has no space at all for free water; its highest possible MC is the FSP, about 30%. Balsa wood represents the other extreme: its cell cavities comprise $^8/_{10}$ of the green volume; water-soaked the species holds over 500% moisture.

As wood of living trees ages, some air penetrates the cell cavities and the moisture content drops below that of water-soaked wood. The actual average MC of *green* sapwood of North American wood species ranges from about 45 to 250%. Sapwood of softwoods generally contains more water than that of hardwoods, because it has the safer water-conducting system of the two, and lower specific gravities on average. In the sapwood of Douglas fir, the species most used, the MC averages almost 150%.

Heartwood of green hardwoods contains in some species more, in others less water than sapwood, the difference generally being small. Taking the average of all native hardwoods, both parts of the tree are equally moist. Not so in softwoods, their heartwood is practically always less wet than sapwood, typically containing half as much water.

3.5. SHRINKING AND SWELLING IN GENERAL

Shrinking and swelling are dimensional changes or movements caused by loss or gain of bound water, respectively. Wood that dries below fiber saturation shrinks; moisture-adsorbing wood swells. Shrinking and swelling cause many problems and have severe consequences, which I shall treat in several sections, restricting this one to general concepts of shrinking and swelling.

Cell Wall Movements

Moisture migrating out of the cell wall in the drying process leaves no empty space behind, rather the wood constituents close in on the water sites and cause the cell wall to contract. Correspondingly, during moisture adsorption the water molecules do not fill empty spaces, instead they make room by squeezing the wood constituents apart and causing the cell wall to expand.

This understanding of the cell wall as an essentially compact, nonporous mass grew out of the fact that moisture-adsorbing cell walls expand, become more permeable, and lose strength, while drying brings about the reverse effects. There is, however, some merit in viewing the cell wall as porous, because the moisture migrates in and out. Researchers even have calculated an internal surface (about 5 million $in.^2/in.^3$ of wall substance) and see the incorporation of moisture as a surface reaction, as adhesion to the internal surface in the form of a thin layer of molecules. Therefore we call the moisture uptake process *adsorption* and not *absorption.* Of course, uptake of free water is *absorption.*

Contracting cell walls can theoretically pull back from the cell cavities, enlarging the cavity space. Conversely, cell walls might theoretically expand and encroach on the cavities, without changing wood's outside dimensions. In reality the cell cavities retain their size and it is the outside dimensions which change: wood shrinks and swells.

These movements, when restricted, develop strong forces. Shrinkage restraint causes drying checks. Wood that swells under restraint exerts a stress as high as that required to compress it (Section 4.1). In flooded hardwood strip flooring, the swelling stress is theoretically high enough to push out walls. Actually, flooded floors will bulge upward in a barrel shape, overcoming the withdrawal resistance of the nails or of the glue bond to the subfloor. The floors bulge rather than push sidewise because the water comes from above and because the edges of the flooring strips are machined to a slight angle for close joints at the surface. The top layer presses more than the bottom layer, and this imbalance in the floor pries it off the subfloor.

Volumetric Shrinking

The cell wall contractions and expansions appear at the wood surface in full size: wood roughly shrinks and swells by the volume ΔV of exchanged bound moisture. The volumetric change can be expressed in terms of exchanged bound moisture ΔMC_b [%] and mass of the ovendry wood W_d [g], on which the MC is based:

$$\Delta V = \frac{\Delta MC_b}{100} \cdot W_d \cdot v\,[cm^3]$$

The v, standing for the specific volume of water [cm^3/g], converts units of mass into units of volume; it can be dropped because it is unity in the metric system. Shrinkage S is the percentagewise change, based on green wood volume $V\,[cm^3]$:

$$S = \Delta V/V \cdot 100 = \Delta MC_b \cdot W_d/V\,[\%]$$

With the specific gravity G in place of the ratio W_d:V follows

$$S = \Delta MC_b \cdot G\,[\%] \tag{1}$$

This is the mathematical form of the initial statement *wood shrinks by the volume of lost bound moisture.*

Most woods shrink less then Eq. 1 implies, some only two thirds of the calculated value S. Equation 1 is derived here mainly to illustrate the relationship between dimensional changes, MC, and specific gravity. Wood technologists rarely calculate shrinkage, working rather with published measured values for each species (Tables 4.1 and 4.2).

Equation 1 is an approximation also because wood does not shrink in exact proportion to the loss of bound moisture (Fig. 3.2). Real pieces start shrinking above fiber saturation, sometimes already around 50% moisture content, when the wood surfaces dries faster than the center and when water tension in cell cavities causes contraction (see Section 3.8).

Swelling stands for the percentagewise change (based on dry wood volume) and is therefore numerically larger than shrinkage S by a factor of $S/(100 - S)$, but it shows the same relationship to moisture and specific gravity, otherwise.

The volumetric shrinkage means a loss for people who dry their own lumber. In Douglas fir the loss averages 6%; many species shrink more, others shrink less. The total of longitudinal, radial, and tangential shrinkage percentages roughly equals volumetric shrinkage, the exact volumetric percentage being slightly smaller for geometric reasons.

3.6 SHRINKAGE IN FIBER DIRECTION

Builders and homeowners pay little attention to volumetric shrinkage; their problem is linear instability in length, thickness, and width of the pieces. Wood shrinks and swells longitudinally, radially, and tangentially by different amounts, transversely much more than longitudinally. Like other properties, the longitudinal-transverse anisotropy results from the orientation of fibers and especially from the longitudinal alignment of cellulose chains. Water

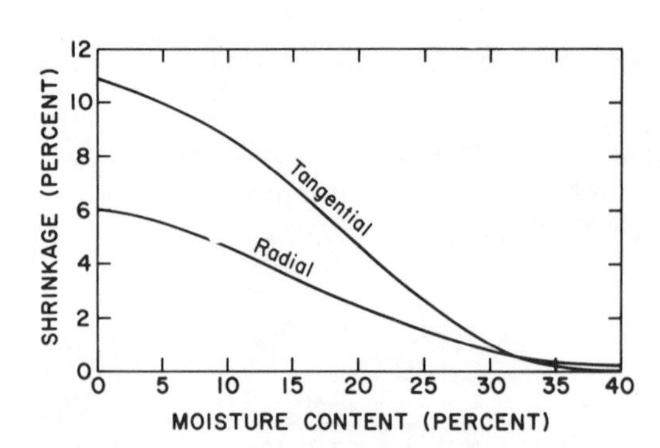

Figure 3.2. Average relation of shrinkage to moisture content of European beech. (By W. C. Stevens.)

penetrating cell walls does not break the chains, but it squeezes bundles of parallel chains farther apart.

Longitudinal shrinkage is negligibly small for most practical purposes and varies percentagewise a good deal within the same species. Some pieces going from green to ovendry shrink more than 0.5%, others only 0.1%, the majority 0.2 to 0.3%. The much higher reaction-wood shrinkage has been discussed. Dense wood shrinks no more than light wood, a fact that should not be surprising, since there is no reason why thick cell walls should shrink lengthwise more than thin walls—just as a thick metal stick contracts in length during cooling no more than a thin stick.

Shrinkage in fiber direction deviates from Eq. 1 also in another respect: many pieces start shrinking not from the fiber saturation point on, proportionally to the loss of bound water, but rather at much lower moisture contents, for example near 15% or even less. Some species start shrinking later than others, but the delay seems to occur in some pieces of all species.

Longitudinal shrinking influences overall dimensions of wood-frame buildings. The height of a house, for example, depends on the length of wall studs, which, because of seasonal fluctuations in temperature and resulting RH fluctuations, shrink and swell in outside walls. The lengthwise shrinking is of the same order of magnitude as thermal expansion of steel. However, temperature variations also cause thermal expansion in wood; this compensates for shrinking so that lumber studs change in length less than steel studs. In overall dimensions wood-frame buildings are essentially stable.

Longitudinal shrinking and swelling cause entrance doors to warp in cold weather. In the door, moisture shifts from the warm inside (house side) skin to the cold outside (street side) skin, for reasons explained in Section 3.10. The inside skin shrinks while the outside skin swells. There results a difference in length of, at most, a few millimeters, but enough for noticeable warpage. As the inside surface curves inward, becoming concave, the door cannot close properly. Most flush doors have plywood skins that swell and shrink more than does solid wood in fiber direction (Section 6.1). These are the main cause of the warpage, but the lumber stiles along the door's vertical edges warp under the moisture shift to some extent too, and restrain the skins very little.

Doors with thick aluminum foil sandwiched into both plywood skins warp less or not at all, but they cost more and seem to be not on the market. Door manufacturers mollify dissatisfied customers with assurances that the door will straighten out in spring. Storm doors reduce warping by raising temperatures at the main door's outside surface. Impervious paint at the inside and a porous finish at the outside assist by keeping house humidity out of the door and letting moisture escape from the outside surface.

The seasonal uplift of roof trusses has essentially the same cause as the warpage of entrance doors. The lower truss chord, deeply buried in insulation, is warmer and drier than the upper chord. Lengthwise swelling of the top chord and shrinking of the lower chord cause the chord as a whole to rise in the center, lifting the ceiling by 1/4 in. or so off the partitions.

3.7. TRANSVERSE SHRINKING AND SWELLING

Transverse shrinkage opens joints in furniture and between flooring strips; it causes loose handles in hammers, wood surface checks, and many other defects. Swelling may have worse consequences, when doors are too wide to close, drawers stick in humid air, and panels bulge. From green to ovendry, even the lightest woods shrink transversely more than 2% (Tables 4.1 and 4.2). In addition, the changes in size differ in the tangential and in the radial direction, a fact which causes special problems that are the main subject of this section.

According to a rule of thumb, the tangential movements exceed the radial movements by a factor of two (Fig. 3.2); the actual ratio varies from species to species. Some woods such as black walnut, mahogany, and yellow birch shrink only about 1.4 times more tangentially than radially; in others, including most oaks, the ratio exceeds two. White pine seems to be worst in this regard with an average near 2.9.

Various explanations have been advanced for the lower radial shrinkage, but none is completely convincing. Very likely, several factors either enlarge tangential shrinking or reduce radial shrinking. The most obvious cause is the restraint provided by the radially oriented cells of rays.

This tendency to shrink tangentially more than radially causes *V-cracks* in pieces that include the pith (Fig. 3.3). The chance for V-cracking increases with the absolute difference between tangential and radial shrinkage, and not with the ratio between them. For example, a species shrinking 8% tangentially and 5% radially cracks more than a *2.5 and 1.0%* species. When one is drying tree disks (*tree rounds*) for use as table tops, one should avoid major differences in moisture content between the various zones, since moisture differences create stress that contributes to cracking. It is much easier of course to avoid moisture differences in thin disks than in long trunks. Similarly, rounds of small diameter dry more uniformly than large disks. Slow drying minimizes the difference in moisture content between surface and center, thus reducing cracking, but it may also lead to one large crack in place of many small ones. Utility poles, logs of log cabins, and thick squared posts with boxed pith cannot be dried slowly enough to prevent cracks, because the center, of some species at least, decays before reaching a safe low moisture content.

In flat-grained lumber the tangential-radial shrinkage anisotropy causes various distortions, most notably *cupping* (Fig. 3.4, top). Since in width direction the bark-side surface comes closer than the pith side to tangential orientation, the boards shrink at the bark side more than at the pith side, and cupping is the result. Strictly speaking, cupping is not restricted to

Figure 3.3. Tree rounds of bur oak with and without *V*-cracks.

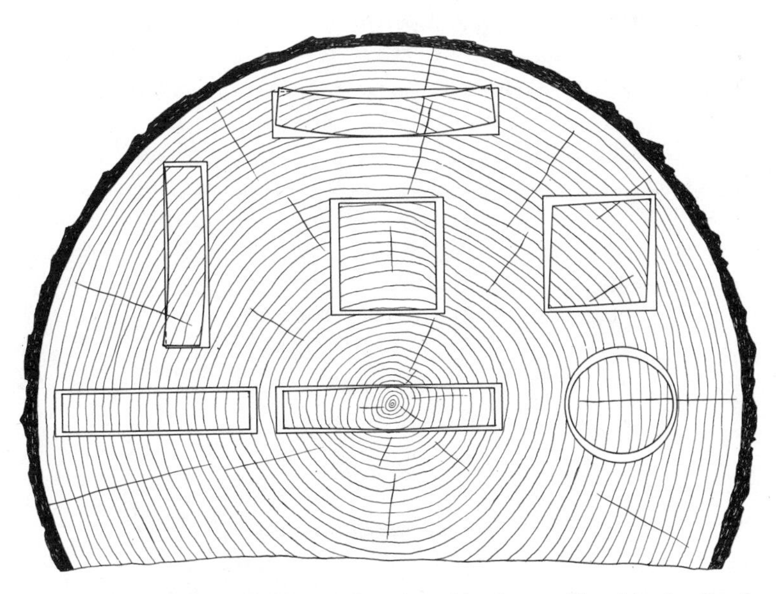

Figure 3.4. Characteristic shrinkage and distortion of lumber as affected by the direction of annual rings. (Photograph courtesy of U.S. Forest Products Laboratory.)

flat-grained lumber. A surface with pith shrinks by the smaller radial amount, while on the opposite surface a narrow flat-grained zone has the larger tangential shrinkage; boards of this kind actually cup most (Fig. 3.5), even though they are mainly edge-grained. Cupping can be a severe defect, especially near butt joints of horizontal siding boards.

All the distortions shown in Fig. 3.4 increase with the tangential-radial anisotropy; again the absolute difference between—rather than the ratio of—tangential and radial shrinkage is what counts.

3.8. CONTROL OF SHRINKING AND SWELLING

Many people consider dimensional instability as the main disadvantage of wood. The competition—metals and plastics—expand and contract from temperature changes less than wood shrinks and swells in the transverse directions. Wood can, however, be stabilized to a great extent in various ways, discussed here.

Drying and Conditioning. Dry the wood to its final moisture content, the EMC of use, before shaping. For outside exposures aim at 12% MC, except in the dry Southwest where 9% is recommended. Since inside EMCs are lower (Fig. 3.6), air-drying in the open is inadequate for interior woodwork. Instead the material must be dried indoors or in kilns, at least during the last stage. If the relative humidity in the workshop or at the factory exceeds that of the final location, the raw material must first be dried to a moisture content well below the final EMC to accomodate for adsorption during manufacturing processes.

In air of certain relative humidity, the wood moisture content approaches the EMC very slowly at decreasing rate (asymptotically). Therefore it is practical to dry at lower RH for a lower MC in the surface layers, and to conclude with a *conditioning* step in which the surface layers readsorb moisture from air of the final RH. The term *conditioning* also refers to any exposure of fairly dry wood to bring its moisture content to the EMC of that air. To condition wall panels to the desired moisture content, lean them against the wall for a few days before nailing; panel the wall when the air is unusually dry to get tight joints.

Drying and conditioning essentially exclude the possibility of large shrink-

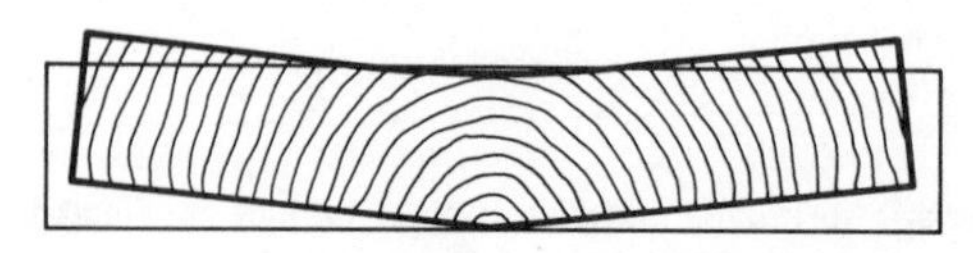

Figure 3.5. Cross section of cupped lumber with the pith at the lower surface. Tangential shrinkage 12%, radial shrinkage 6%, width 12 in., thickness 2 in. (By N. N. Hsu and R. C. Tang.)

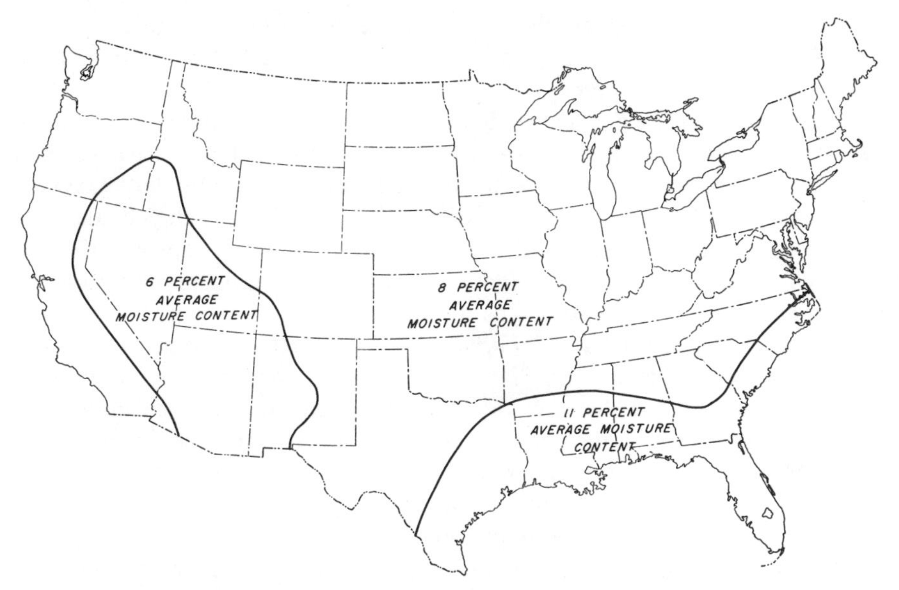

Figure 3.6. Recommended average moisture content for interior use of wood products in various areas of the United States. (Photograph courtesy of U.S. Forest Products Laboratory.)

age, but to reduce swelling and shrinking due to changes in humidity while the product is in use, apply one of the following principles.

Orientation of Fibers and Annual Rings. Orient the fibers and annual rings for lowest dimensional changes in the crucial direction. The panel door is a good example of proper fiber orientation; its frame of vertical stiles and horizontal rails around thin panels assures that the door shrinks in length and width only by the minimal longitudinal amount. The panels fit into frame grooves, in which they may shrink and swell freely without affecting the door's length and width.

Proper orientation of annual rings can minimize size-changes of many products, among them flooring strips and siding boards. Edge-grain flooring shrinks in width less than the flat-grain variety and retains tighter joints between strips. As siding, edge-grain lumber rarely cups, and its low shrinkage is essential for the durability of paint (Section 14.3).

Crossbanding. Crossbanding means *bonding of layers with fiber directions at right angles to each other to the effect of mutual restraint of movements.* The principle is explained in Section 6.1 for plywood, the simplest and best known crossbanded product. Plywood shrinks in the plane only slightly more than lumber in fiber direction.

The pieces of fiberboard and mat-formed particleboard lie in the plane of the panel like the layers of plywood, but with fiber directions oriented at random within the plane rather than at right angles. Therefore the mutual restraint is slightly less than in plywood.

Of course, the crossbanding concerns the plane only. In thickness the panels shrink and swell at least as much as would the isolated components. For more detail see Chapter 6.

Exclusion of Liquid Water. Rain water or condensation water, which rapidly seeps into the cell cavities of unprotected wood, takes weeks or even months to evaporate; it causes excessive moisture fluctuations, large average MCs, and sometimes decay. The penetration should be excluded and can be, through overhanging roofs, for example, and with water-repellent finishes (Section 14.8). Rabbets below window panes and end faces of siding boards particularly need to be made water-repellent.

Retarding the Exchange of Vapor. Finishes such as varnish and paint retard moisture exchanges and dampen the fluctuations, but do not prevent them. The pores in finishes are wide enough for passage of vapor molecules, though in most cases, fortunately, not wide enough for the much larger molecule aggregates of liquid water. For more information about moisture-excluding effectiveness of finishes see Section 14.2.

Many surfaces of wood products are covered with *overlays*, either in the form of thin sheets of plastic (*contact films* and others) or of *laminates* (Section 14.9). Overlays retard moisture much more than traditional finishes, which have pores from solvents, but overlays still let some vapor pass. Only sheets of metal are vapor-tight. For most purposes, however, the overlays seal satisfactorily, since they also reduce swelling and shrinking through mechanical restraint.

Stabilizing with Chemicals. Certain chemicals incorporated in wood reduce shrinking to various degrees, depending mainly on the location of the chemical. In cell cavities the wood preservative creosote and other oils serve as an internal coating that retards drying and moisture absorption, in this way reducing the fluctuations in moisture content.

For real stabilization—beyond retardation of swelling and shrinking—chemicals have to penetrate the cell wall, where they work either by cross-linking or as a bulking agent. *Cross-linking* agents tie cell wall components together in the shrunken state and close the sites to water. This treatment requires only a small percent of formaldehyde, for example, but it embrittles wood and is rarely worth the limited stabilization achieved.

Bulking agents replace bound moisture and keep the wood swollen. One kind, synthetic resin, penetrates the cell wall as a liquid and hardens under heat or by radiation. Salts, including common salt, are easier to apply, but they stabilize less; as hygroscopic substances they cause the wood to be damp in humid air, and they cause corrosion of fasteners. Sugar attracts wood-destroying organisms and, like salts, stabilizes only partially.

By far the most popular bulking agent for wood is the *carbowax* polyethylene glycol 1000 (PEG). For treatment, immerse green wood into the aqueous PEG solution for a few weeks or so, depending on wood thickness and permeability; the relatively large PEG molecules diffuse through water into the cell walls to replace bound moisture. Heat accelerates the penetration. Later in air the free water and remainders of bound moisture evaporate. Properly treated PEG wood shrinks and swells very little or not at all. This method has found widespread use for carvings, turnings, tree rounds, and other novelty items. The *Crane Creek Company*, P.O. Box 5553, Madison, Wis. 53705, sells PEG in convenient quantities and provides further information free of charge.

3.9. DRYING OF LUMBER

This section explains principles of drying, essential kiln techniques, and how to dry without special equipment; it is supposed to serve hobbyists who buy trees to be sawn in small mills and then dry the lumber themselves. Builders may want to recognize drying defects and to learn the principles of drying, because lumber on the market typically has up to 19% moisture content, too much for flooring, cabinets, and other interior woodwork. [The *American Softwood Lumber Standard PS 20-70* (Section 5.1) defines *dry* lumber as *lumber that has been seasoned or dried to a MC of 19% or less*; lumber that contains more moisture is *green*].

In the drying process, free water evaporates near the wood surface out of water-filled capillaries, whose menisci then pull moisture from the core, causing rapid flow to the evaporation front. As the capillaries become empty, the evaporation front recedes to deeper layers, from which moisture migrates by diffusion at decreasing rates to the surface. Eventually moisture diffuses all the way—as bound moisture across cell walls, as vapor across cell cavities and through pits—to the wood surface, driven by the higher vapor pressure in deeper, moister layers. Essentially wood dries simply on exposure to dry air. The difficulty, besides achieving the right final moisture content, lies in avoiding drying defects, which are discussed here in their sequence of occurrence.

Discolorations

In the first stage of drying, growth of fungi may discolor and even decompose the wood. The fungi need water and oxygen (as explained in Section 10.1). In waterlogged wood they have no oxygen, while below 20% MC they lack water; but inbetween they thrive. The wood should rapidly pass through this critical range, under good ventilation of all surfaces, before these damaging organisms proliferate. If for some reason fast drying is not possible, brush the lumber with a fungicide. Of course, high temperatures kill the fungi also.

Some discolorations result from various abiological chemical reactions of extraneous wood constituents with impurities at the lumber surface or with air. Like the fungal changes in color, such reactions require water as a solvent and do not occur below 20% MC. These obviously were the reasons for choosing

19% MC as the upper limit of the *dry* condition in the *American Softwood Lumber Standard.* At 19% or less, lumber can be stored and transported in compact piles—without stickers between the layers—for unlimited time.

End Splits and Surface Checks

Since moisture travels longitudinally many times faster than transversely, end faces dry more rapidly than lateral surfaces and tend to shrink earlier. Adjoining wood restrains the shrinkage, causing transverse tension and finally end splits (Fig. 3.7). Covers of plastic film or canvas over the ends and end coats of paraffin or asphalt-base paint retard the end-grain drying and prevent or at least reduce splitting.

Likewise, the surface of lumber dries faster than the core and shrinks earlier. The core prevents the surface shell from shrinking in width, brings it under tension stress, causing surface checks (Fig. 3.7). Flat-grain boards check the most because restraint is in the direction of largest shrinkage (tangential), and because light earlywood bands near the surface restrain the larger shrinkage of dense latewood.

Surface checking is caused obviously by large differences in moisture content between shell and core. The surface should dry not much faster than moisture flows from the core to the shell, meaning slow drying, which in turn favors discolorations. The compromise is to first dry speedily, until most free water has evaporated and the dry-looking surface begins to shrink. Then dry at slow rates in humid air. Only later, when the core's MC has dropped below fiber saturation, should RH decrease to the desired EMC, or slightly

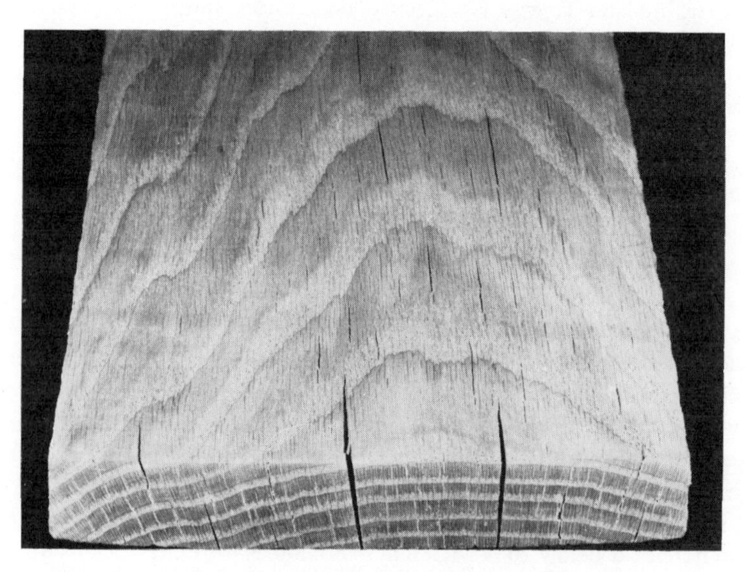

Figure 3.7. Oak plank showing surface checking and end splitting.

less, until the lumber has reached the final MC. This regime also helps reduce splitting.

Wood-wise hobbyists, for example, who receive green boards early in spring, will stack the material outside in the shade for a few days or weeks. Then they transfer the stack to a cool basement, leaving it there all summer and during the following heating season. The basement air, humid at first, gradually becomes dry in fall when temperatures outside drop below the temperature of the heated basement. Early in February basement humidity reaches a minimum and brings the wood shell's MC far down, while the still moist core continues to dry. When in March RH gradually increases, the shell readsorbs some moisture, so that by the end of the heating season the boards are uniformly dry. To prevent thick lumber of the wood of dense species from drying too fast the first winter, cover the stack with plastic film or transfer it to an unheated shed.

Internal Stresses, Warp, and Casehardening

Drying stresses are in themselves a defect. In the internally stressed lumber, tension forces in surface layers are in equilibrium with compression forces in the core. Lengthwise sawcuts disturb the equilibrium. The two halves obtained by ripping an internally stressed two-by-four, for example, do not remain straight but bow. Planing off one surface layer disturbs the equilibrium likewise, the opposite surface layer pulls the piece over and causes warp: the board cups from transverse stress and bows from longitudinal stress.

The stresses considered so far result directly from differences in moisture content between the layers and disappear as soon as the moisture equalizes. Strong stresses however may exceed the proportional limit, causing plastic stretching and compression, leaving the wood under internal stress even after moisture equalization. A copper wire tension-stressed almost to the point of failure illustrates this plastic stretching, remaining elongated after removal of force. Likewise, blows of a hammer against the head of a bolt permanently shorten the bolt shaft. Drying stresses stretch lumber surfaces and compress the core widthwise and lengthwise in similar ways. Dimensions of surface and core actually may remain the same, but the wood has what is called *sets*, which are explained in more detail in Section 12.4.

Because of sets, lumber which has dried in stacks between misaligned sticks and is therefore bowed (Fig. 3.8), retains the bow when unstacked. Similar sets result from one-sided drying; besides bowing they cause other kinds of warp such as twist and crook. Each piece should be ventilated equally well at opposite surfaces. On the other hand, set may be used to reduce warp when compensating for one-sided shrinkage in flat grain-, knotty-, and reaction-wood boards, and to this end the pieces should be held straight in weighted stacks.

Lumber internally stressed by sets is characterized by the misnomer *casehardened*, which originates from metallurgy where it means a *hard*

Figure 3.8. Warp of stacked lumber due to misaligned stickers. (Photograph courtesy of U.S. Forest Products Laboratory.)

surface. When planed one-sided or ripped, casehardened lumber warps permanently. Since heat plasticizes (Section 12.1), casehardening occurs more in kiln drying than in air-drying. On the other hand, high temperatures can reverse and relieve casehardening. For relief, the kiln operator either applies steam briefly to induce swelling stress in the wood surface and reverse the stretching, or he/she conditions the uniformly dry lumber at temperatures sufficiently high for relaxation of all stresses.

Honeycombing and Collapse

The overstretched surface of casehardened lumber restrains shrinkage of the core, causes tension in the core, and contributes to internal checks (*honeycombing*, Fig. 3.9). The main cause of honeycombing is believed to be cell collapse from water tension (Section 1.5) developed by menisci in tiny capillaries. Wood cells collapse infrequently and only in portions of the piece. It is collapse within the core beneath an intact shell that causes honeycombing, while collapse of earlywood or latewood (Fig. 3.10) in edge-grain boards produces corrugated surfaces or *washboarding*, named after the appearance of old-fashioned washboards.

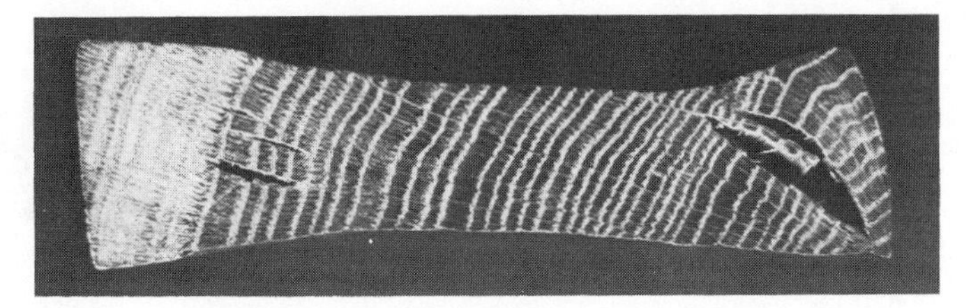

Figure 3.9. Cross section of dried oak board, showing that heartwood collapse caused either honeycombing or excessive shrinkage.

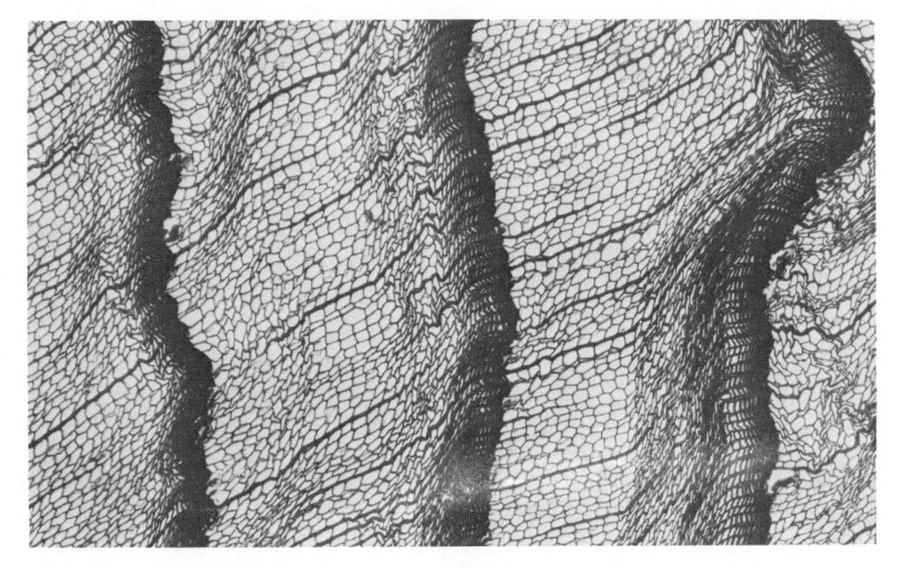

Figure 3.10. Cross section microphotograph of collapsed western red cedar. (Photograph courtesy of U.S. Forest Products Laboratory.)

Drying Rate

How fast lumber dries depends on the particular condition and on the material. One-inch softwood boards protected from rain dry outside in summer within one or two months. Inside, without ventilation and without heating, the process lasts longer. Dense species such as oak require more drying time than softwoods; in the green state they contain less moisture percentagewise but more mass; in their smaller cavities, moisture travels slowly. Besides, as a result of the large shrinkage, dense species must be dried slowly to prevent checking and splitting. Firewood, in which checks are no problem and which can be exposed to very dry air immediately, dries within one summer in any species, provided the pieces are short, well ventilated at the ends, and protected from rain.

Dry-kilns are heated chambers equipped for controlling temperature, humidity, and air movement. The control according to proved schedules of temperature and humidity permits fast drying with less defects than in air-drying. Mainly due to high temperatures and the resulting great differences in vapor pressure, kiln drying of softwood boards takes only a few days. Most lumber sold by dealers has been kiln dried at the sawmill.

3.10. VARIATION AND ACCUMULATION OF MOISTURE IN HOMES

Wood is protected from rain indoors, and people rarely let water stand at the wood surface. One might also suppose that humidity fluctuates less inside then outside. In fact, however, wood in buildings may stay wet for months in some sections of homes where humidity stays high, while elsewhere the air appears very dry. Why this is so, where moisture accumulates, and how to prevent detrimental conditions is the subject of this section.

Cause and Consequences of Humidity Fluctuations

Two simple facts of physics determine the distribution of moisture in buildings. The first fact results from diffusion of vapor molecules into all accessible spaces: in equilibrium the air throughout a house contains same amounts of vapor per unit volume, *the vapor has the same pressure everywhere.* The second fact, as previously explained: *warm air can hold more vapor than cold air.*

Figure 3.11 illustrates the consequences. This home presumably has reached a state of equilibrium in which, according to the first fact, vapor pressure is the same in all rooms. The condition at left prevails during warm July, the condition at right in January. In both cases the humidity in the living quarters is assumed to be typical for the time of the year, while humidities in other spaces are calculated as follows for the basement, taking vapor pressures at saturation (p_s) from tables in engineering handbooks to determine the actual partial vapor pressure p (Section 3.1):

p_s at 77°F, temperature in living quarters	0.4590 psi
p at 77°F and 65% RH, $0.65 \cdot 0.4590$	0.2984 psi
p_s at 66°F, temperature in basement	0.3164 psi
RH in basement, $0.2984/0.3164 \cdot 100$	94%
EMC for 94% RH, according to Fig. 3.1	23%

The damp basement air favors growth of mold on wood surfaces, while internal wood moisture is sufficiently high for decay. Fortunately, moisture content drops during the heating season, but this drop in turn involves much shrinkage.

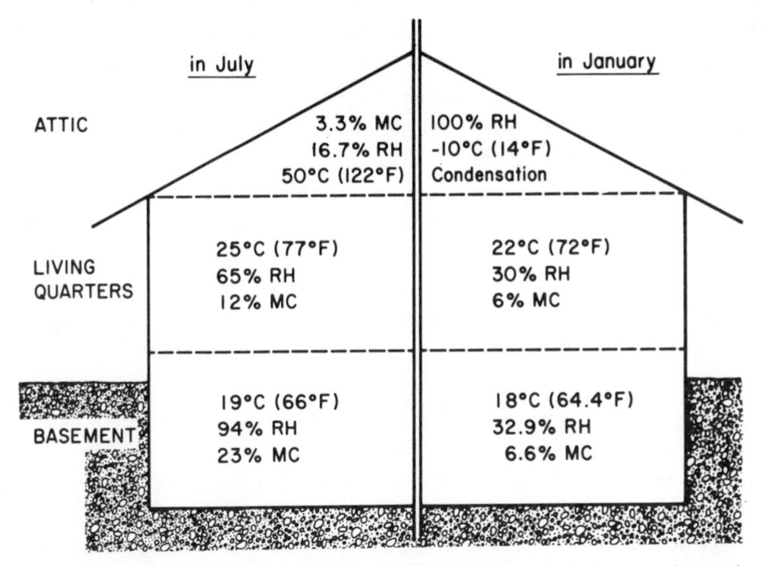

Figure 3.11. Humidities and equilibrium wood moisture contents in a ranch-type home.

In the attic, humidity and moisture content vary even more than in the basement in this example. However, most attics are separated from living quarters and have their own independent ventilation to the outside, so an equilibrium of moisture content with the rest of the house may not be expected.

How to Avoid Dampness and Condensation

Any heat generated, be it pumped out of a refrigerator or produced in other appliances, lowers relative humidity by raising the temperature. Forced air circulation throughout the home holds the RH in humid rooms and corners down, again by raising the temperature, without actually removing any moisture. Electrical dehumidifiers work in two ways: by condensing water out of the air and by generation of heat. Air conditioners condense moisture when they cool air to below the dew point, that is when operating at low temperatures and at low air speed.

Due to the second fact, in winter, air is dry in warm but not in cold rooms or cool corners. Closets at poorly insulated outside walls stay cool and consequently damp, even if nobody hangs up wet clothes; open closet doors reduce the relative humidity by letting in heat. Water condensation on cold window panes illustrates that temperature is crucial for the humidity.

Inside exterior walls the temperatures drop so low that vapor may condense and form thick chunks of ice, which engulf insulation blankets and conduct plenty of heat to the outside. In spring the ice melts, water soaks into the wood to damage paint, and runs out of the wall base. Avoid the condensation by blocking vapor's path to the cold zone with impervious wallpaper or with a vapor barrier such as the polyethylene film *Visqueen* (a trade name) stapled to the studs before wall panels are installed.

Attic spaces can be protected in the same way with a vapor barrier in the ceiling of living quarters. Outside ventilation through louvers under overhanging roofs and in gabel ends serves the same purpose. In addition, ventilation holds the attic heat down in summer when the sun heats up the roof. Ventilation is also more reliable: vapor barriers, besides still having some permeability, become punctured during construction or else let vapor pass along their edges. The main benefit of vapor barriers in ceilings is that they retain the humidity in the living quarters in winter on a higher and more comfortable level.

Wood as Air Conditioner

Seasonal fluctuations of wood moisture in homes have a significant beneficial effect of which few people are aware. As wood adsorbs moisture when the humidity rises in summer and gives off moisture into dry air in winter, it helps equalize fluctuations in humidity, moderating the uncomfortable dry and humid peaks. Wood does this both seasonally and daily, equalizing the changes from day to day or day to night; it conditions the air in homes.

To what extent wood *air conditions* depends on many factors. The amounts of exchanged wood moisture can be impressive. In average single-family wood-frame homes which have roughly 10,000 lb of wood inside the vapor barrier, a difference in moisture content of 10% between January and July amounts to 1000 lb or 120 gallons of water. Common humidifiers and de-humidifiers exchange roughly the same amount. One cord of dry oak firewood stacked in the basement will adsorb in summer close to 40 gal. This equalizing effect of wood is probably one of the reasons why man likes wood.

4

Strength and Elasticity of Wood

In structures wood carries loads without breaking and without noticeable deformation. In diving boards it should deflect under the weight of the diver, while in baseball bats the impact-deformation seems not to matter. This chapter deals with both qualities, with the ability to resist loads without failure, and with deformations caused by loads. Included is the influence of growth characteristics and of environmental conditions on strength, but not design values for structural lumber, which appear in Chapter 5.

4.1. KINDS OF STRENGTH

This section explains the mechanical properties listed in Tables 4.1 and 4.2, enabling the reader to evaluate the various wood species and take full advantage of their wide variety. I shall show where different kinds of strength play a role and explain the tests in which the values of Tables 4.1 and 4.2 were determined.

Tension Parallel to Grain

Tension is first discussed because of its simplicity; few structural members in wood-frame houses are tension-stressed. Figure 4.1 illustrates a rare example in the chord of the roof truss: loads of snow and wind of the roof tend to push the two rafter ends and the walls outward; the rafters *pull* the chord, which in turn ties the structure together and prevents the house from collapsing.

Table 4.1. Properties[a] Parallel and Perpendicular to Grain of Some Woods Grown in the United States

Official Common Tree Name	Spec. Gravity[b]	Bending Str.	Mod. of Elas.[c]	Work Max. Load[d]	Compres. Par.	Compres. Per.[e]	Shear Str. Par.	Tension Per.	Side Hardness[f]	Rad. Shrinkage[g]	Tang. Shrinkage[g]
					HARDWOODS						
Alder, red	0.41	9,800	1.38	8.4	5820	440	1080	420	590	4.4	7.3
Ash, white	0.60	15,400	1.74	17.6	7410	1160	1950	940	1320	4.9	7.8
Aspen, quaking	0.38	8,400	1.18	7.6	4250	370	850	260	350	3.5	6.7
Basswood, American	0.37	8,700	1.46	7.2	4730	370	990	350	410	6.6	9.3
Beech, American	0.64	14,900	1.72	15.1	7300	1010	2010	1010	1300	5.5	11.9
Birch:											
Sweet	0.65	16,900	2.17	18.0	8540	1080	2240	950	1470	6.5	9.0
Yellow	0.62	16,600	2.01	20.8	8170	970	1880	920	1260	7.3	9.5
Butternut	0.38	8,100	1.18	8.2	5110	460	1170	440	490	3.4	6.4
Cherry, black	0.50	12,300	1.49	11.4	7110	690	1700	560	950	3.7	7.1
Cottonwood, eastern	0.40	8,500	1.37	7.4	4910	380	930	580	430	3.9	9.2
Elm:											
American	0.50	11,800	1.34	13.0	5520	690	1510	660	830	4.2	7.2
Rock	0.63	14,800	1.54	19.2	7050	1230	1920	---	1320	4.8	8.1
Hickory, shagbark	0.72	20,200	2.16	25.8	9210	1760	2430	---	---	7.0	10.5
Honeylocust	0.64	14,700	1.63	13.3	7500	1840	2250	900	1580	4.2	6.6
Magnolia, southern	0.50	11,200	1.40	12.8	5460	860	1530	740	1020	5.4	6.6
Maple, sugar	0.63	15,800	1.83	16.5	7830	1470	2330	---	1450	4.8	9.9
Oak:											
Live	0.88	18,400	1.98	18.9	8900	2840	2660	---	---	6.6	9.5
Northern red	0.63	14,300	1.82	14.5	6760	1010	1780	800	1290	4.0	8.6
White	0.68	15,200	1.78	14.8	7440	1070	2000	800	1360	5.6	10.5

Sassafras	0.46	9,000	1.12	8.7	4760	850	1240	---	---	4.0	6.2
Sweetgum	0.52	12,500	1.64	11.9	6320	620	1600	760	850	5.3	10.2
Sycamore	0.49	10,000	1.42	8.5	5380	700	1470	720	770	5.0	8.4
Tupelo, black	0.50	9,600	1.20	6.2	5520	930	1340	500	810	5.1	8.7
Walnut, black	0.51	14,600	1.68	10.7	7580	1010	1370	690	1010	5.5	7.8
Yellow poplar	0.42	10,100	1.58	8.8	5540	500	1190	540	540	4.6	8.2
				SOFTWOODS							
Baldcypress	0.46	10,600	1.44	8.2	6360	730	1000	270	510	3.8	6.2
Cedar, western red-	0.32	7,500	1.11	5.8	4560	460	990	220	350	2.4	5.0
Douglas-fir, coast	0.48	12,400	1.95	9.9	7240	800	1130	340	710	4.8	7.6
Fir, white	0.39	9,800	1.49	7.2	5810	530	1100	300	480	3.3	7.0
Hemlock, western	0.45	11,300	1.64	8.3	7110	550	1250	340	540	4.2	7.8
Larch, western	0.52	13,100	1.87	12.6	7640	930	1360	430	830	4.5	9.1
Pine:											
Eastern white	0.35	8,600	1.24	6.8	4800	440	900	310	380	2.1	6.1
Loblolly	0.51	12,800	1.79	10.4	7130	790	1390	470	690	4.8	7.4
Longleaf	0.59	14,500	1.98	11.8	8470	960	1510	470	870	5.1	7.5
Ponderosa	0.40	9,400	1.29	7.1	5320	580	1130	420	460	3.9	6.2
Shortleaf	0.51	13,100	1.75	11.0	7270	820	1390	470	690	4.6	7.7
Redwood, old-growth	0.40	10,000	1.34	6.9	6150	700	940	240	480	2.6	4.4
Spruce, Engelmann	0.35	9,300	1.30	6.4	4480	410	1200	350	390	3.8	7.1

[a] From Wood Handbook; properties of clear straight-grained specimens; strengths (psi) and elasticity at 12% moisture content.
[b] Specific gravity based on ovendry weight and volume at 12% moisture content.
[c] Modulus of elasticity determined in bending test (million psi).
[d] Work to maximum load in static bending test (in. $lb/in.^3$).
[e] Compression perpendicular to grain, stress at proportional limit (psi).
[f] Side hardness (lb).
[g] Shrinkage from green to ovendry (%).

Table 4.2. Properties[a] Parallel and Perpendicular to Grain of Some Woods Imported into the United States

Common Name and Botanical Name of Species	Spec. Gravity[b]	Bending Str.	Mod. of Elas.[c]	Work Max. Load[d]	Compres. Par.	Shear Str. Par.	Side Hardness[e]	Rad. Shrinkage[f]	Tang. Shrinkage[f]
Apitong (Dipterocarpus spp.)	0.65	16,200	2.35	---	8,540	1690	1200	5.2	10.9
Balsa (Ochroma pyramidale)	0.17	2,800	0.55	---	1,700	300	100	3.0	7.6
Banak (Virola koschnyi)	0.48	10,800	1.72	8.1	5,720	1300	640	4.6	8.8
Khaya (Khaya ivorensis)	0.46	10,700	1.39	8.3	6,460	1500	830	4.1	5.8
Lauan:									
Almon (Shorea almon)	0.44	11,300	1.67	---	5,750	1090	590	3.8	8.0
Red (Shorea negrosensis)	0.47	11,300	1.63	---	5,890	1220	680		
Lignumvitae (Guaiacum sanctum)	1.09	---	---	---	11,400	---	4500	5.6	9.3
Mahagony (Swietina macrophylla)	0.48	11,600	1.51	7.9	6,630	1290	810	3.7	5.1
Obeche (Triplochiton scleroxylon)	0.35	7,500	0.86	6.9	3,930	990	430	3.1	5.3
Ramin (Gonystylus bancanus)	0.64	18,400	2.17	17.0	10,080	1514	1300	3.9	8.7
Rosewood (Dalbergia latifolia)	0.86	16,900	1.78	13.1	9,220	2090	2630	---	---
Spanish-cedar (Cedrela odorata)	0.36	7,900	1.01	5.6	4,450	---	500	4.1	6.3
Teak (Tectona grandis)	0.63	12,800	1.59	10.1	7,110	1480	1030	2.2	4.0

[a] From Wood Handbook; properties of clear straight-grained specimens; strengths (psi) and elasticity at 12% moisture content.
[b] Specific gravity based on ovendry weight and volume at 12% moisture content.
[c] Modulus of elasticity determined in bending test (million psi).
[d] Work to maximum load in static bending test (in. lb/in.^{3}).
[e] Side hardness (lb).
[f] Shrinkage from green to ovendry (%).

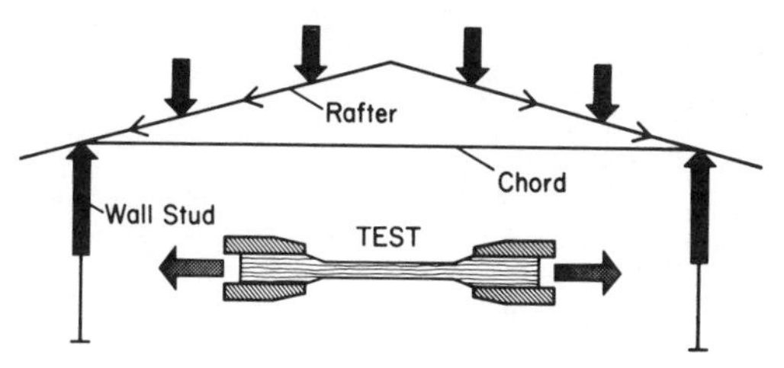

Figure 4.1. Tension in fiber direction; and forces in the frame of a house showing how tension in the chord of roof trusses comes about.

In the *small clear* test specimen of Fig. 4.1 the forces produce *stress* that depends on the cross section area:

$$\text{stress} = \frac{\text{force}}{\text{cross section}}$$

In the British system of units, stress is expressed in lb/in.^{2} (psi).

Each tensile stress, no matter how small, stretches the material. If we pull a rubber band, it expands, but returns to the original length once that force is discontinued, showing *elastic* behavior. Wood is elastic like rubber, only its elongations are so small they go unnoticed without sensitive instruments to detect them.

Since elongation depends on the length of the specimen, we express the change in length with the relative measure *strain:*

$$\text{strain} = \frac{\text{elongation}}{\text{length}}$$

The unit of strain, length per length (in./in.), is omitted. A 50 in. stick stretched to 50.5 in., for example has a strain of 0.01 or 1%.

Each stress increment increases the strain. In wood, as in most materials, strain increases at the rate of stress (Fig. 4.2). The ratio of *stress-to-strain* is constant; it tells how strongly the material resists deformation and how much deformation is caused by certain stress, in short show stiff the material is. Engineers call this ratio *modulus of elasticity* (MOE):

$$\text{modulus of elasticity} = \frac{\text{stress}}{\text{strain}}$$

The MOE is expressed in the unit of stress (psi). Rubber has a low MOE, responding to stress with large strain, whereas steel's MOE is high compared with that of wood (Fig. 4.2).

A proportionality between stress and strain exists in many materials only up to a certain degree of stress, called the *proportional limit* (*P* in´Figure

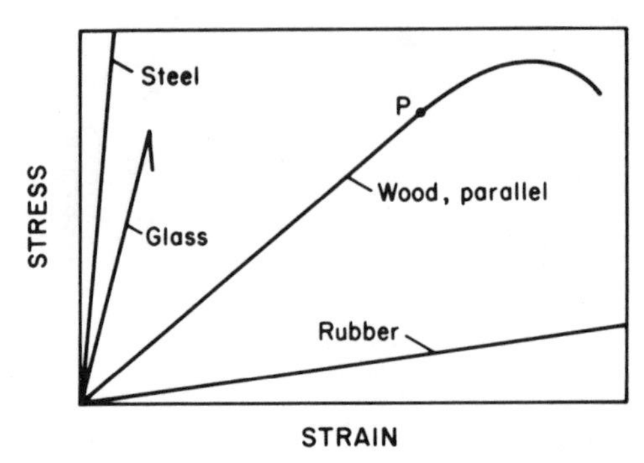

Figure 4.2. Relation between stress and strain, four different materials.

4.2). The limits of steel and rubber fall beyond the range of this graph. Glass has no proportional limit; it is *brittle,* meaning it *breaks abruptly at small strain*.

The maximum stress that the material reaches is called *strength.* In this narrow sense *strength* means the *ultimate stress at which the material fails*. In a broader sense *strength* includes resistance against deformation; a material which scarcely deforms when loaded (having a high MOE) is *strong* in this sense, irrespective to its ultimate stress.

Compression Parallel to Grain

Studs in walls (Fig. 4.3), posts under girders, and legs of chairs are under compressive stress parallel to their grain. In the test shown, the increasing load first reduces the specimen's length unnoticeably, pressing all tiny components slightly closer together in fiber direction. Again the MOE is calculated from the observed strain, which in this case is defined as the *negative elonga-*

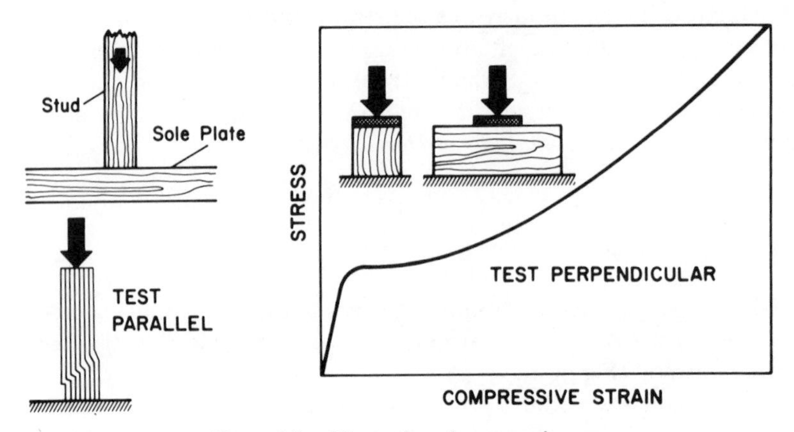

Figure 4.3. Illustration of compression.

tion-to-length or *compression-to-length* ratio. At the point of ultimate stress (*compressive strength*), fibers buckle suddenly along a diagonal line and the specimen may split apart in fiber direction, losing almost all ability to resist the load.

Columns whose length exceeds 11 times their thickness will bend to the side and buckle as a whole before compression failure appears. They cannot be designed considering compressive strength alone, rather their *length-to-thickness* ratio and MOE must be considered in determining allowable stress.

Compression Perpendicular to Grain

Sole plates under studs in walls (Fig. 4.3) resist load perpendicular to their grain. In the corresponding test, increasing loads compress the specimen at first in proportion to the stress, as shown in the lower part of the stress-strain graph in Fig. 4.3; later the cell walls give way without failure in an elastic fashion. Upon load removal the cell walls recover, and the specimen regains its original thickness.

From about 1% compressive strain on up, some cell walls will crush and bulge into cell cavities, yielding a more rapid compression. As these cells collapse, the wood becomes denser (Section 12.4) and more resistant; further loads cause smaller and smaller compression increments. The wood carries unlimited loads; no ultimate stress appears and so *strength* in the narrow sense cannot be defined.

Collapse of all cells would reduce the thickness of sole plates in most wood species to less than one half and the wall would sag. To prevent noticeable sagging, building codes restrict the design stresses in compression perpendicular to grain to a stress below the proportional limit.

Static Bending

Bending forces occur in many structural members of buildings and in numerous other products. Examples are joists under floors (Fig. 4.4), roof sheathing, beams of bridges, bows for shooting arrows, and bed frames. To understand the nature of bending, consider the exaggerated model in Fig. 4.4. In the original square piece all four lateral surfaces have the same length, but after bending the concave upper surface is much shorter than the convex lower surface. Bending compresses the concave half while stretching and tension-stressing the convex half; it is a combination of tension and compression. Between the two halves, along the dashed *neutral* plane, the length remains unchanged, there the stress is zero.

The strains and stresses are highest (*extreme*) at the upper and lower surfaces, where the material fails when bent. *Bending strength,* also called *modulus of rupture* (MOR), lies somewhere between compressive and tensile strength.

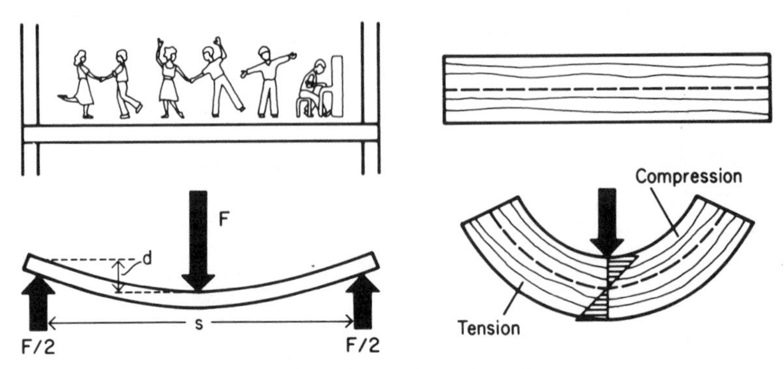

Figure 4.4. Illustration of bending.

Engineers do not measure the extreme stress, they rather calculate it on the basis of the assumed linear stress distribution from force F, span s, specimen depth or height h, and specimen width w (Fig. 4.4). The MOR is the calculated extreme stress (psi) caused by the ultimate load F':

$$\text{MOR} = \frac{3\,F's}{2\,wh^2}$$

The extreme stress permitted in structures appears in *design value tables* under the designation *allowable stress in bending* or as *extreme fiber in bending* (Table 5.1).

The deflection d (Fig. 4.4) involves lengthwise compressions and elongations. Stiffness in fiber direction and bending deflection are obviously closely related. On this basis engineers calculate the MOE from deflections in bending tests. In fact, the MOE of wood is usually determined by bending, and not by tension and compression, where measuring strain requires very sensitive instruments. The deflection depends on span s and depth h in the third power:

$$d = \frac{Fs^3}{4\,\text{MOE}\,wh^3}$$

The equation indicates something well known, that short deep beams deflect hardly at all. Due to the strong influence of depth, a ruler bends over its flat face much easier than over its edge, and beams several times deeper than wide provide high bending stiffness with a minimum of material.

Dynamic Bending

Under *static* bending the load increases gradually, whereas in *dynamic* bending it shoots high in fractions of a second, as in hammer handles, baseball bats, diving boards. Dynamic bending properties are measured either in impact bending tests or in toughness tests. In the *impact* test, a weight released successively from increasing heights falls on the stick until the stick fails, or until it deflects by a certain large amount. The final height from which the weight drops indicates the impact strength in units of length (in.). In the *toughness* test, a pendulum hammer bends and breaks the stick in a single blow. The energy (in. · lb) consumed in breaking the stick serves as the measure of toughness. To enable comparison of data from tests of sticks of various sizes, toughness may be expressed in units of energy per cross section areas (in. · lb/in.^{3}).

Impact strength and toughness are *dynamic* strengths (*strength* in the broad sense), which differ entirely from the modulus of rupture determined in static bending tests. Dynamic strength is a quantity of energy instead of an ultimate stress like the MOR. In dynamic bending the integral product of force and deflection counts, whereas the MOR depends on maximum force only. *Work to maximum load* (Tables 4.1 and 4.2), the integral product of force and deflection measured in static bending tests, correlates very closely with impact strength and toughness; it makes a good measure of dynamic strength.

Hickory surpasses all other native woods in dynamic bending strength, because of its high ultimate stress and because it deflects greatly before failing—it is tough wood. Hickory's high ultimate stress is explained by high density, its large deflection by low lignin content. Brittle wood may be higher in maximum force and MOR but much lower in dynamic strength, as illustrated in Fig. 4.5. Tough wood splinters when breaking, in contrast to brash wood that breaks in fairly even surfaces at small deflections, and very suddenly, the most frequent cause of excessive brashness being incipient decay (Section 10.3).

Bending strength, stiffness in bending, and dynamic bending strength are termed *flexural* properties.

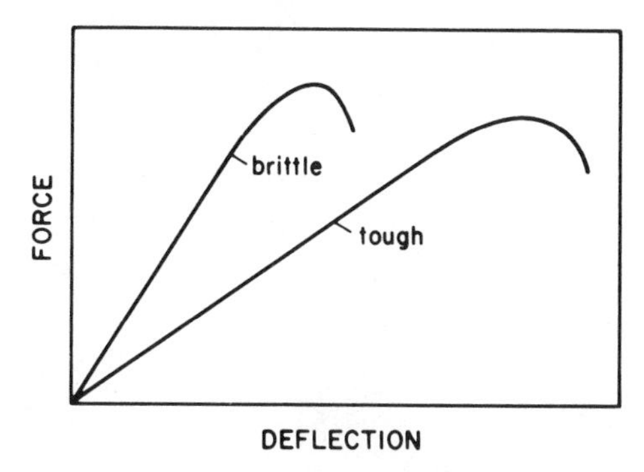

Figure 4.5. Effect of force on deflection of brittle and of tough material.

Shear

Shear stress causes layers to slide over each other. The term comes from *shearing,* that is, *cutting with shears* or large scissors whose edges move against each other, forcing the material's layers apart (Fig. 4.6). The tool edges are not sharp but straight and they approach each other in the same plane.

Ladder rungs represent an example of shear-stressed wood; they are of course also bending-stressed, but shear becomes crucial when a climber puts weight next to the sidepiece, as indicated for the lower rung of Fig. 4.6. This is *transverse* shear, also called *vertical* shear, because most transverse shear results in structures from vertical loads. Wood rarely fails under transverse shear, so data of this kind are not readily available.

Shear parallel to grain has more significance. Mortise and tenon joints fail when flanks of the tenons shear off along the grain. Parallel shear determines the strength of many glue joints, as well as of joints made with timber connectors, screws, bolts, and dowels. The ultimate stress in shear parallel to grain is determined on shear block specimens (Fig. 4.6); the edge of the upper bar descends slightly offset rather than directly above the edge of the lower bar. Without this offset, we would be measuring compressive rather than shear strength whenever the fibers are oriented as in the figure.

Parallel shear determines the load-carrying capacity of many beams as follows. Consider boards stacked one above the other but not fastened together (Fig. 4.6). Under load, the boards deflect as shown, slipping against each other so that their flush ends become staggered. Glued boards in laminated beams cannot slip but tend to, and so do imagined layers of solid beams. The tendency to slip causes parallel shear, here called *horizontal shear* because of horizontal position of beams. In deep beams the shear stress is highest near the ends where no adjoining wood prevents the layers from slipping; deep beams of short length may fail in shear strength rather than in tensile and compressive strength.

The test block of Fig. 4.6 rests on the end-grain surface; and the force acts in fiber direction. If the end-grain surface is rotated 90° to face the viewer,

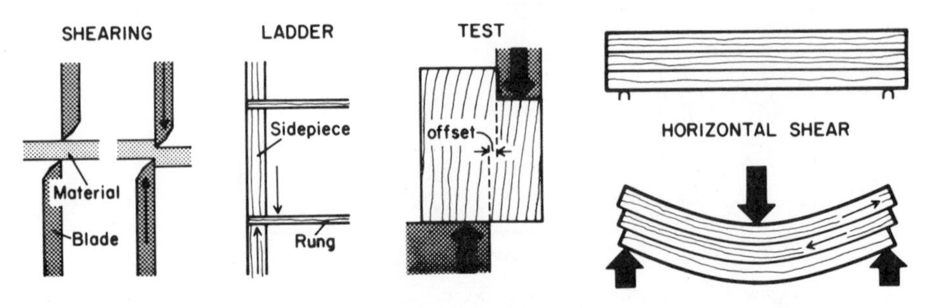

Figure 4.6. Illustration of shear.

the force will press perpendicular to the fibers, which roll over each other in *rolling shear*. The resistance against rolling shear is only about one quarter as great as parallel-shear strength.

Hardness

The spike heels, which were fashionable around 1960, dented and ruined many wood floors. Spike heels disappeared but other indentations keep occurring—telephones dropping on the floor, children pounding their toys on it. There are also scratches, for example from sliding chairs. The damage is due to insufficient *hardness*, which is defined as *resistance against indentation.*

The hardness values of Tables 4.1 and 4.2 are determined by pressing a steel ball with a surface of 4 cm^2 into the wood surface. Hardness is measured by the load needed to force penetration to one half of the ball's diameter. The ball pushes cell walls aside and forward into the cavities of adjoining cells. How much force this requires mainly depends on the *wall-to-cavity* ratio, as does specific gravity. Being closely related to density and specific gravity, hardness makes a good indicator of many other wood properties. Specific gravity gives better estimates, but for pieces too heavy to lift, hardness is easier to estimate, for example by pressing a fingernail into the surface.

Side hardness at lateral surfaces and the *end hardness* of end-grain surfaces differ slightly. End faces are in most but not in all species harder than the sides.

4.2. EFFECT OF GROWTH CHARACTERISTICS

Wood strength has still other variables; some (Section 4.3) result from service conditions and can be influenced, while those discussed in this section lie in the wood itself and are beyond control in the final product.

Orientation of Fibers

Fiber orientation has the greatest influence on tension, where strength parallel to the surface of the trunk is 15 to 50 times higher than transverse strength, or about 20 times higher, taking the average of all species (Fig. 4.7). In compression the ratio of *longitudinal-to-transverse strength* varies between 3 : 1 for moist wood and 15 : 1 for brash, ovendry material. Only in a few extremely dense species from tropical forests are the transverse and longitudinal compressive strengths about identical. Bending, as a combination of tension and compression, naturally falls between the two in the ratio. For other properties of strength the *longitudinal-to-transverse* ratio averages 10:1.

In intermediate directions strength changes with the angle, as illustrated in Fig. 4.7. Surprisingly low strength in a piece is mostly caused by fiber deviation—slope of grain, cross grain. In tension, slopes of *1 in 25* can already

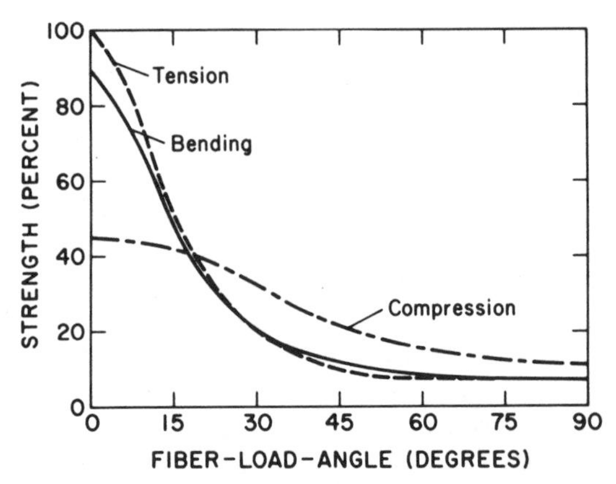

Figure 4.7. Effect of fiber-load-angle on strength of clear wood.

have significant effects, while in compression a substantial drop begins at slopes of *1 in 10*. Dynamic bending strength is more sensitive in this regard than static bending strength, which explains why cross-grained handles of hammers and axes break so easily.

In whole tree stems deviations have relatively little influence on strength because the fibers spiral around the stem and do not *run out of the piece,* always transferring the load to adjoining fibers above and below. That is why small stems are frequently stronger than round sticks of lumber the same size, having the same cross grain even though the stems consist of juvenile wood.

Orientation of Annual Rings

Under transverse compression (Fig. 4.8) and tension, strength depends on the direction of annual rings and rays. Their influence varies among the species and is small compared with the influence of the fiber direction. Rays increase strength in the radial direction (Fig. 4.8*a*), while dense latewood bands strengthen in the tangential direction (Fig. 4.8*c*). Diffuse-porous woods, which by definition lack dense latewood bands, are up to two times stronger radially than tangentially. In softwoods where tiny rays occupy small portions of the volume, the influence of latewood bands dominates over or at least equals

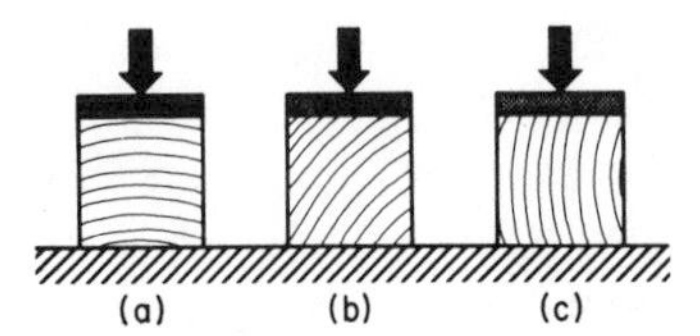

Figure 4.8. Radial (*a*), intermediate (*b*), and tangential (*c*) compression.

the influence of rays; in this group, species with distinct latewood are more than 1.5 times stronger tangentially than radially. Drying checks, being quite common in flat grain but rare in edge-grain lumber, weaken the wood particularly in tangential tension.

In the diagonal direction (Fig. 4.8*b*), dense latewood bands of softwoods and also of ring-porous hardwoods slide along porous earlywood, with the effect that wood is weakest diagonally.

The orientation of annual rings has some influence in bending of deep short beams which fail in horizontal shear. Porous earlywood layers have relatively little strength and are sheared apart if oriented as in Fig. 4.8*a*, whereas in the orientation of Fig. 4.8*c* the strong latewood bands carry the brunt of the horizontal shear stress.

Specific Gravity

Since the cell wall substance of all wood species is almost identical, wood strength increases with the amount of that substance, that is with increasing specific gravity. Within species the relationship between strength and specific gravity is roughly linear. For example, a piece of 0.6 specific gravity exceeds the strength of a piece with specific gravity 0.4 by a factor of approximately 0.6:0.4 or 1.5, rather more than less. This rule of thumb applies to compression, tension, static bending, shear, and hardness. Dynamic bending strength increases more steeply than specific gravity, in a power function; a piece twice as dense for example may be four times higher in dynamic strength.

From species to species and when comparing sapwood and heartwood, differences in minor wood constituents modify these relationships, but these rules still serve as a crude guide.

4.3 EFFECT OF SERVICE CONDITIONS

This section deals with the influence of humidity, temperature, and duration of load-bearing. Humidity is more important than temperature except in the case of fire.

Influence of Moisture

Water molecules, entering dry cell walls and squeezing cell wall components apart, reduce the cohesion between components and weaken the wood. This bound moisture may also be seen as an internal lubrication for the separation of wood components under stress. Since free moisture affects strength only in the form of ice, strength increases during drying when the moisture content drops below the fiber saturation point. The increase varies from species to species, between the various kinds of strength, and from piece to piece. Figure 4.9 shows the average relationship.

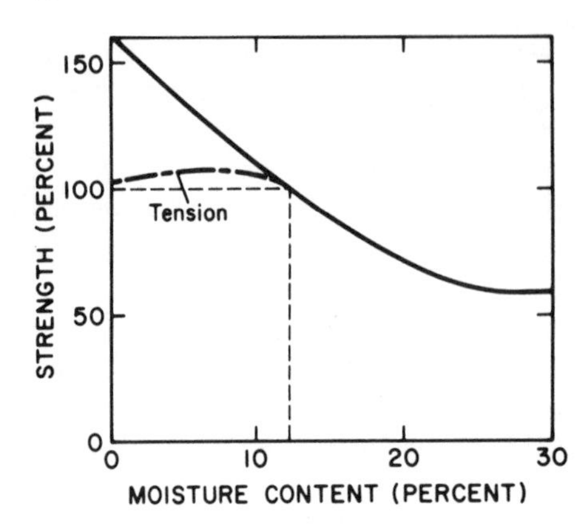

Figure 4.9. Effect of moisture content on strength at room temperature.

Strength increases generally during drying all the way down to the ovendry state, except for tensile strength and the bending MOE, which peak at around 5% moisture content. These peaks are due to water molecules first linking wood hydroxyl groups together, before additional water separates wood components.

The depicted strength-moisture relationship applies to wood which dried very slowly. Drying checks naturally weaken wood and compensate for the strength increase; badly checked dry lumber may be weaker than green lumber.

Dynamic bending strength is only slightly affected by the moisture content, being often lowest at around 15% MC. The influence of moisture on the maximum force and on the ultimate deflection counteract each other. It takes more force to break dry wood, but moist wood bends farther. (Hence the Boy Scout rule for kindling, *if you can't snap it, scrap it.*)

Temporary Influence of Temperature

The vibrations of molecules and atoms in solids intensify as the temperature rises. The intenser vibration pushes wood molecules farther apart and causes thermal expansion. Farther apart the molecules cohere less, wood loses strength (Fig. 4.10). This effect of heat is immediate and reversible. When cooled the molecules move closer together again, wood contracts and regains strength.

Moist wood reacts to heat more than dry wood, because in the moist swollen state the components are farther apart from the very beginning. High temperatures affect the ultimate stress more than the modulus of elasticity. Strength declines less longitudinally than transversely,because longitudinally the distance between molecules changes less inasmuch as thermal expansion is less in that direction (Section 8.1). Dynamic bending strength sometimes increases, sometimes decreases with rising temperature, as increased ultimate deflection counteracts the decreased cohesion.

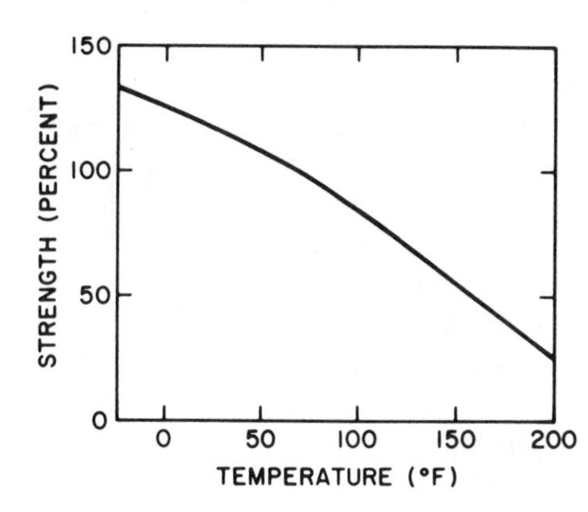

Figure 4.10. Temporary effect of heat on strength at 12% moisture content.

Cooling, of course, has the opposite effect, it strengthens wood. In moist material frost adds additional strength (especially compressive strength) through ice in cell cavities. Frozen wet wood of low specific gravity, with ice-packed cell cavities comprising most of its space, resists at the freezing point several times more compression stress than unfrozen wood at slightly higher temperature. Freezing temperatures have no consequences for bound moisture, which is in a solid state at all temperatures, also above the freezing point.

Air temperature changes tend to trigger humidity changes in reverse: in hot zones of buildings the air is dry, in cold zones humid. Wood's dryness at higher temperatures counteracts the immediate weakening effect of heat, so that in use wood strength changes less than indicated in Fig. 4.10.

Permanent Effect of Heat

Excessive heat weakens wood permanently. At high temperatures the molecular vibrations are so strong that wood constituents disintegrate. Thermal degradation begins in dry wood around 100°C (212°F), while in moist wood a sort of hydrolysis causes degradation from about 70°C (158°F) upward, becoming noticeable after several days. For more detail about thermal degradation see Section 8.2.

As mentioned in Section 3.2, the chemical changes involve reduced equilibrium moisture content. When exposed to an atmosphere of 65% relative humidity, for example, lumber which has been dried at temperatures above 100°C (212°F), attains a moisture content of about 11% and not the usual 12%. The corresponding gain in strength makes up for thermal degradation, so that high-temperature-dried and steamed wood are practically as strong as wood dried at low temperature, except in dynamic bending. Dynamic-bending strength suffers significantly from temperatures above 100°C (212°F), because heat embrittles wood and reduces its ability to bend.

Duration of Stress

Bending loads cause immediate deflections, which seem to remain constant. However, time shows that under heavy constant load the deflection gradually increases; the material *creeps.* Roof ridges of old improperly designed buildings sag between the supporting rafters. Old floors may settle in the center of their span so much that door frames are distorted and gaps open up along interior walls. Creep, the gradual increase in deformation under a constant load, occurs not only in bending but under all kinds of stress. Loaded columns slowly become shorter, tension chords stretch in roof trusses. Like elastic strain, fortunately, creep due to compression and tension (but not in bending) is so small that only fine instruments detect it.

The extent of creep depends on the load and on duration of loading. Small bending stress causes no noticeable creep even when sustained over hundreds of years, while under very heavy loads, creep appears in a few minutes. The rate of creep changes with time, being relatively fast at first then slowing down until failure starts. A beam may creep in seven years following its first year no more than it did during the first year.

Creep indicates internal changes, whose nature is discussed in Section 12.1. Here suffice it to note that the load first forces some particles to slip into new positions where they equalize the stress. Later the cohesion between some particles breaks, and adjacent particles absorb the additional load, carrying it for some time. When later they fail too, still other neighbors must acquire the substantial additional load. At this point the rate of creep accelerates and failure of the whole piece is imminent.

Creep has significant consequences. Stresses which wood can resist for many years are less than 60% of the stress required to bring it under in a few minutes. Man instinctively knows that prolonged loading increases the chance of failure: over a weak bridge he moves with deliberate speed, particularly avoiding going back and forth.

Frequent loading and unloading promotes internal changes; hence rapid loading cycles weaken wood faster than does constant stress, though the latter is higher when averaged over time. Even worse are cycles of tension and compression, as in bending alternately to opposite sides—just as wire breaks when bent back and forth. Engineers describe the progressive damage from repeated loading with the term *fatigue*; in *fatigue tests* they load and unload the material in thousands of cycles. *Fatigue strength* is the stress that can be applied a specific number of times before failure; for an infinite number the fatigue strength of wood amounts to only 30% of short-time strength.

5

Lumber, Glulams, and Veneer

This chapter and Chapter 6 deal with *intermediate* wood products, which are processed further for use in the construction of homes, or to build furniture, or baseball bats, or many other *final* products. The three kinds covered here retain all the natural features of wood and involve less processing than do panels as discussed in Chapter 6. Lumber is by far the most important among the three intermediate products. Glulams (short for *structural glued laminated timbers*) are made from lumber for construction. Veneer is most used in plywood and for that reason is usually treated together with plywood in the literature. I include veneer here because it has all of wood's natural properties.

5.1 GRADING AND MEASUREMENT OF LUMBER

Lumber is defined as *the product of the saw and planing mill, not further manufactured than by sawing, resawing, passing lengthwise through a standard planing machine, crosscutting to length, and matching.* Moldings such as hand rails, quarter rounds, and base boards along the floor line still belong to this category of *lumber*, whereas window frames and stairs are *millwork*.

Grading in General

Like many commodities, lumber is marketed in various grades, which enable buyers to choose the quality they wish, using short terms such as *No. 1 Common.* Unfortunately the number of lumber grades is almost endless. The grades vary not only with wood species, with defects like knots, with forms of manufacture and dimensions, but also with the use category: such as *structural joists* and *planks.* Furthermore, each sawmill and each dealer are legally free to create and define any grade. A few facts, however put this apparent monster into order and simplify the grading.

First, most sawmills grade softwood lumber according to rules of regional lumber associations (such as the Western Wood Products Association) and stamp the association grade sign on each piece. The associations define the grades in handy grading rule booklets, supervise the grading through inspectors in the mills, and provide reinspection services to resolve grading disputes over shipments. The system works fairly well because associations and mills have vital interests in the integrity of their grades as well as in satisfied customers.

Legally the associations can define and name grades as they please. One association may grade certain lumber *No. 3,* while another prefers the less negative term *Standard.* It is the *American Softwood Lumber Standard* that provides common ground between them all, and for hardwood the National Hardwood Lumber Association issues uniform *Rules* for the whole nation. Large consumers, such as the U. S. Defense Supply Agency and major wholesalers, buy only lumber that conforms with these two regulations, thus forcing the industry to comply.

The American Softwood Lumber Standard

This regulation was issued in 1970 by the National Bureau of Standards (NBS) of the U. S. Department of Commerce under the designation *PS 20-70* in the series of *NBS Voluntary Product* Standards. (Prior to 1970 there had been *Commercial Standards* or CS.) *PS 20-70* differentiates between *yard lumber, structural lumber,* and *factory and shop lumber.* Each of the three categories has its particular grading principles.

Retailers sell mainly *yard lumber,* that is, lumber for general purposes, usually simply called *lumber.* The standard defines two classes of yard lumber: *Select*—characterized by *good appearance and finishing qualities*—and the lower class *Common—suited for general construction and utility purposes.* Both classes are then divided into groups and subgroups, for example, the *Select* class into one group that is *suitable for natural finishes* and another group *suitable for paint finishes.*

Structural lumber is intended for use in construction where each piece must provide certain minimal resistance against mechanical stresses. In this category only such features as affect strength reduce the grade. Mineral stains, for example, which downgrade *yard lumber,* are irrelevant in *structural lumber.* A typical use of the *factory and shop lumber* category is in mass-produced windows and doors; its grading resembles that of hardwood (see Hardwood Rules).

The standard is a guide for drafting grading rules and not a grading rule in itself, leaving definition of grades to the lumber associations. Therefore grades do vary somewhat from one association to another. The NBS and its American Lumber Standard Committee do not control or enforce application of *PS 20*-70; the standard becomes legally binding only in conjunction with building codes, purchase orders, and sales contracts.

Hardwood Rules

The *Rules for the Measurement and Inspection of Hardwood and Cypress Lumber* do define numerous grades of the various hardwood species, including some imported woods and lumber for very special uses, such as *hard maple heel stock.* Eight standard grades serve as skeleton:

Firsts	Selects	No. 2 Common	No. 3A Common
Seconds	No. 1 Common	Sound Wormy	No. 3B Common

Firsts and *Seconds* are usually combined as one grade under the abbreviation *FAS.* The *Rules* include cypress lumber of the softwood tree baldcypress because some cypress lumber is used like hardwood; when used like other softwoods, cypress grading follows the *American Softwood Lumber Standard.* Similarly, some hardwoods are graded like softwoods, the best known example being aspen for studs, a typical softwood-use category; otherwise the mills grade aspen acccording to the *Rules* for hardwood. Most hardwood lumber on the U. S. market is graded by these *Rules*, excepting only some *finish* products. Flooring for example is covered by a Product Standard of the NBS.

Grading Principles and Buying by Grades

Hardwood and softwood grading differ in principle for good reason. Most softwood lumber is used by carpenters, who install each stud and each beam as it comes, disregarding all knots and other undesirable features, whereas hardwood traditionally serves as the material of cabinetmakers, who cut out imperfections and use the clear wood.

The softwood grade depends on the effect of imperfections on strength, workability, and appearance. By contrast, the hardwood grade depends on how much clear material the piece contains. A long softwood board with a single knot in the center, for example, falls into a low grade, whereas in hardwood the knot matters little since it leaves nearly the whole board clear. Softwood *factory and shop lumber* makes the exception, intended as it is for cut-up purposes, and should therefore be graded by the percentage available in cuttings of sufficient quality.

The occasional buyer of a few pieces explains his or her needs to the retailer, probably naming the species but not knowing the grade, and the salesperson then chooses among available material. For small purchases a grade rule booklet is not worth the effort, especially as one's retailer carries only a few grades anyway. Commercial buyers of large quantities study the grade brochures and booklets of the various lumber associations, choose the appropriate grade, compare offers before buying; and for buyers of this quantity the grade is available.

Sizes, Measurement, and Definitions

Both the *Lumber Standard* and the *Hardwood Rules* name standard thicknesses and widths for the various groups and use categories. Softwood boards for example can be 1, 1 1/4, and 1 1/2 in. thick. These are nominal dimensions of rough green lumber. According to the *Standard*, when dressed or surfaced (which generally means *planed*) the 1 in. green board, for example, may be only 25/32 in. and the dry board 3/4 in. thick. Customers who order 1 in. boards often notice with disappointment that they get far less than 1 in. Nominal, dressed green, and dressed dry sizes also appear in face widths. Unlike the *Lumber Standard*, the *Hardwood Rules* do not list dry sizes; their surfaced thicknesses and widths are larger for most nominal sizes than in the Standard.

Mills and dealers sell lumber in actual lengths, since longitudinal shrinkage is negligibly small. Standard lengths are multiples of 1 ft in hardwoods and multiples of 1 ft or 2 ft in softwoods.

In North America, the *board measure* is in common use for lumber, a volume measurement whose basic unit is the *board foot*. A piece 1 ft long, 1 ft wide, and 1 in. thick has a volume of 1 board foot (bd. ft.). Obtain the number of board feet in a piece by multiplying the thickness in inches by the width in feet by the length in feet.

As an example, a 16-ft long two-by-six measures $2 \times 1/2 \times 16$, or 16 bd. ft. Nominal size counts, not dressed size. Lumber less than 1 in. in nominal thickness is considered 1 in. thick.

The lumber trade applies numerous abbreviations not only in wholesale but also in retail trade. Measurement in lineal feet, for example, is usually abbreviated *LFT*. The *American Softwood Lumber Standard* lists abbreviations of this sort and defines many terms, among them the following:

Boards. Lumber less than 2 in. in nominal thickness and 2 or more in. in nominal width. Boards less than 6 in. in nominal width may be classified as *strips*.

Dimension. Lumber from 2 in. to, but not including, 5 in. in nominal thickness, and 2 or more in. in nominal width. Dimension may be classified as *framing, joists, planks, rafters, studs, small timber,* and so forth.

Timbers. Lumber 5 or more in. nominally in its smaller dimension. Timber may be classified as *beams, stringers, posts, caps, sills, girders, purlins,* and so forth.

Rough Lumber. Lumber that has not been dressed (surfaced) but has been sawed, edged, and trimmed, at least to the extent of showing saw marks in the wood on four longitudinal surfaces of each piece for its overall length.

Dressed (surfaced) Lumber. Lumber that has been dressed by a planing machine (for purposes of attaining smooth surface and uniform size) on one side (*S1S*), two sides (*S2S*), one edge (*S1E*), two edges (*S2E*), or a combination of sides and edges (*S1S1E, S1S2E, S2S1E, S4S*).

Worked Lumber. Lumber that in addition to being dressed, has been matched (tongued and grooved), shiplapped (edge dressed to make a lapped joint), or patterned (Fig. 5.1).

5.2. DESIGN VALUES FOR STRUCTURAL LUMBER

Roof truss members, wall studs, beams and joists which support floors and ceilings, are typical uses of *structural* or *stress-grade lumber;* that is, *lumber 2 in. or more in nominal thickness for carrying loads, graded solely for its ability to resist stress.* Each stress grade has a *design value* (*allowable stress, working stress*) assigned to it. In balsam fir, for example, 2 to 4 in. thick and 2 to 4 in. wide, *No. 3* of the Northern Lumber Manufacturers Association, at 19% moisture content, the allowable bending stress is 500 psi, the tensile stress 300 psi, and the compressive stress perpendicular to grain 170 psi.

Engineers and architects design structures on basis of allowable stress, calculating how thick the piece has to be and designating this grade in their drawing, so that builders and lumber dealers know exactly which wood grade to select. The National Forest Products Association, as the umbrella of the regional lumber associations, compiled all available allowable stresses for the convenience of users in *Design Values for Wood Construction* (see Bibliography and Table 5.1). The stress grade designations and descriptions are almost identical throughout the United States, differing from association to association much less than does *yard lumber.*.

The *design values* in Table 5.1 fall far short of the strengths in Table 4.1. For example, in the best grade of southern yellow pine *extreme fiber in bending* is 2000 psi, whereas the corresponding *bending strength* of lobolly pine, the weakest among the southern pines, is 12,800 psi. The MOEs in Table 5.1 are about the same as those in Table 4.1; in calculating safe loads on long columns however, these MOEs have to be reduced to one third.

Lumber associations determine the *design values* by means of a two-step procedure, which is described in the standard *ASTM D 245* (see Bibliography). In step one the known average *strength* of small clear samples (Table 4.1) is reduced to the *basic stress* of clear lumber, taking into account the natural

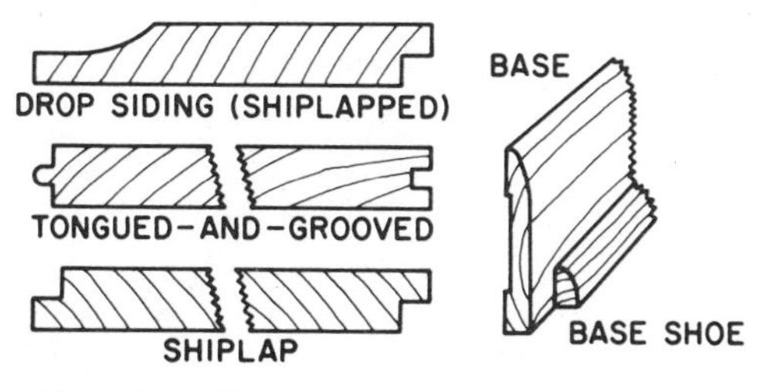

Figure 5.1. Frequent patterns of worked lumber.

Table 5.1. Section of the National Forest Products Association (NFPA) Table *Design Values for Visually Graded Structural Lumber*. (Photograph courtesy of NFPA.)

Special and commercial grade	Size classification	Design values in pounds per square inch							Grading rules agency
		Extreme fiber in bending "F_b"		Tension parallel to grain "F_t"	Horizontal shear "F_v"	Compression perpendicular to grain "$F_{c\perp}$"	Compression parallel to grain "F_c"	Modulus of elasticity "E"	
		Single-member uses	Repetitive-member uses						
SOUTHERN PINE (Surfaced dry. Used at 19% max. m.c.)									
Select Structural	2" to 4" thick 2" to 4" wide	2000	2300	1150	100	405	1550	1,700,000	SPIB
Dense Select Structural		2350	2700	1350	100	475	1800	1,800,000	
No. 1		1700	1950	1000	100	405	1250	1,700,000	
No. 1 Dense		2000	2300	1150	100	475	1450	1,800,000	
No. 2		1400	1650	825	90	405	975	1,600,000	
No. 2 Dense		1650	1900	975	90	475	1150	1,600,000	
No. 3		775	900	450	90	405	575	1,400,000	
No. 3 Dense		925	1050	525	90	475	675	1,500,000	
Stud		775	900	450	90	405	575	1,400,000	
Construction	2" to 4" thick 4" wide	1000	1150	600	100	405	1100	1,400,000	
Standard		575	675	350	90	405	900	1,400,000	
Utility		275	300	150	90	405	575	1,400,000	
Select Structural	2" to 4" thick 5" and wider	1750	2000	1150	90	405	1350	1,700,000	
Dense Select Structural		2050	2350	1300	90	475	1600	1,800,000	
No. 1		1450	1700	975	90	405	1250	1,700,000	
No. 1 Dense		1700	2000	1150	90	475	1450	1,800,000	
No. 2		1200	1400	625	90	405	1000	1,600,000	
No. 2 Dense		1400	1650	725	90	475	1200	1,600,000	
No. 3		700	800	350	90	405	625	1,400,000	
No. 3 Dense		825	925	425	90	475	725	1,500,000	
Stud		725	850	350	90	405	625	1,400,000	
Dense Standard Decking	2" to 4" thick 2" and wider Decking	2000	2300	––	––	475	––	1,800,000	
Select Decking		1400	1650	––	––	405	––	1,600,000	
Dense Select Decking		1650	1900	––	––	475	––	1,600,000	
Commercial Decking		1400	1650	––	––	405	––	1,600,000	
Dense Commercial Decking		1650	1900	––	––	475	––	1,600,000	
Dense Structural 86	2" to 4" thick	2600	3000	1750	155	475	2000	1,800,000	
Dense Structural 72		2200	2550	1450	130	475	1650	1,800,000	
Dense Structural 65		2000	2300	1300	115	475	1500	1,800,000	

variation from piece to piece, long-time service, moisture contents, drying defects, high temperatures, and a safety factor besides. The second step has to do with strength-reducing defects such as knots, and leads from the *basic stress* of clear lumber to the *design value* of imperfect lumber, among the various grades.

The associations describe grades with permissible imperfections and submit the stresses to the National Grading Rule Committee of the American Lumber Standards Committee. The Grading Rule Committee consists of representatives of lumber associations, building officials, building code agencies, home builders, civil engineers, architects, consumers, U. S. Federal Housing Administration, U. S. Defense Supply Agency, U. S. Forest Products Laboratory, and the National Bureau of Standards. The Committee examines the proposed allowable stresses and endorses safe grades. In this way the public and competing associations control grade-drafting.

Most structural lumber is graded visually like the other lumber categories, the grader looking at each piece and evaluating imperfections to assign the grade. A few large sawmills grade by using a machine which bends the piece, registers the resulting deflection, determines the grade based on computed stiffness, then stamps the piece accordingly. Machine grading is based on the fact that stiff pieces tend to be strong, or in other words, that stiffness and strength are statistically interdependent. Design values of machine stress-rated pieces differ in many cases from those of visual grading; in some cases visual grading is higher, in others machine-grading allows more stress.

5.3. STRUCTURAL GLUED LAMINATED TIMBER

Structural glued laminated timber (*glulam*) consists of lumber laminations bonded with an adhesive so that the grain of all laminations runs parallel with the long direction. Glulam is intended—as its full name implies—for structural use and was developed as a stress-rated product, whose manufacture, inspection, testing, and marking follows the *Product Standard 56-73.* At present glulam laminations cannot be veneer, though veneer would give essentially the same properties and is used in *laminated veneer lumber* (*parallel laminated veneer*) for trusses. The term *laminated wood* comprises both *glulam* and *laminated veneer lumber.* In *built-up timber* the layers consist of lumber arranged in parallel, as in glulam, but are joined together with nails or other mechanical fasteners, which connect the layers at points only and provide limited resistance against slippage of layers (see Fig. 4.6, *horizontal shear*). Built-up timber is therefore weaker and less stiff than glulam.

Inherent Characteristics

Consisting of parallel lumber laminations, glulam has essentially the properties of lumber: it is strong lengthwise and weak transversely, swelling and shrinking considerably in thickness and in width but practically not at all in length. Since the bonding requires labor, adhesive, and special facilities, glulam costs on the average two times more than lumber.

Why then do designers and builders choose glulam in place of lumber? Laminated two-by-fours and other small pieces are not used. Glulams are needed for large dimensions not available in lumber. The number of laminations is unrestricted, and individual laminations can be joined endwise as well as edgewise to any desired size. Glulams 130 ft long and 10 ft high have been made. Manufactured sizes are limited not by the raw material but by transportation to the construction site. Very long laminated timbers are therefore made and transported in short sections to be connected endwise with mechanical fasteners onsite.

Glulams have a second main advantage: they can be curved in permanent bends (Fig. 5.2). Manufacturers assemble straight glue-coated laminations and bend the assembly into the desired shape. During bending the laminations slip along each other, as indicated by the staggered ends in Fig. 4.6. If the bending force were removed at this moment, the laminations would slip back and straighten out, but an adhesive bond prevents the recovery.

Figure 5.2. Glulam arch footbridge at Wentworth-by-the-Sea in Newcastle, New Hampshire. (Photograph courtesy of American Institute of Timber Construction.)

The individual layers of glulam are typically 1.5 in. thick; the standard allows at most 2 in. The timber cross section may vary over the length and can be adapted as needed for strength and appearance. For thick parts the manufacturer simply adds short laminations; similarly, laminations of various widths give cross sections of any desired shape. The staggered surfaces, which result, can be planed even.

Besides carrying load, glulams determine the structure's appearance and add to its beauty; they are therefore not only rated for stress but graded by appearance. *PS 56-73* identifies three appearance grades: *industrial, architectural,* and *premium.* The three differ in permitted visible defects such as loose knots, knot holes, knot size, voids, areas of insufficient surfacing, color and grain of laminations, eased corners, and so forth.

Strength and Moisture Relations

Manufacturers use stress-graded lumber—mostly Douglas fir and southern yellow pine, or durable oak for boat timbers—calculating allowable stresses on basis of the lumber's design value. Laminations in nonstructural parts of the timbers need not be stress-graded.

Basically glulam is not stronger than lumber since gluing has no effect on strength, but with wisely selected laminations the bending strength exceeds that of the lumber. Manufacturers place high grade laminations near the surface where bending stresses are highest (Section 4.1) and use low grade pieces for the center near the neutral plane (Fig. 5.3).

In curved timbers the bending induces stress in each lamination and lowers the strength of the timber as a whole. The effect becomes stronger (for obvious reasons) as lamination thickness and curvature increase.

Other strength reductions have to be made for end joints within individual laminations. The joints weaken the tension flank of bending members more than the compression flank. All joints must be of the scarf or at least of the finger type (Section 13.3). Butt end joints hardly transmit tension forces and should

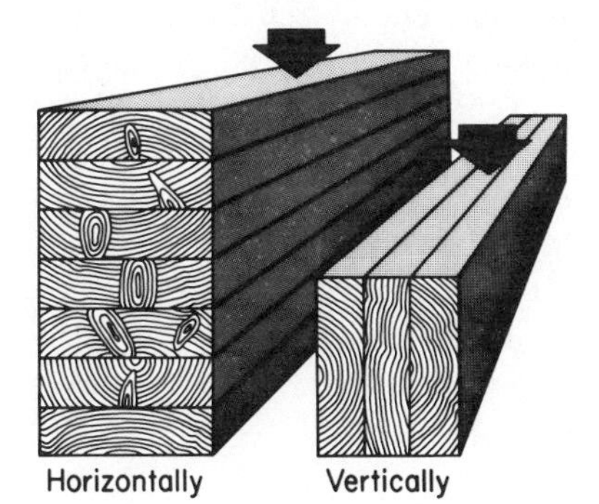

Figure 5.3. Orientation of laminations in glulams. (By R. C. Moody.)

not be included. The strength-reducing joints and knots have to be staggered; dispersion of imperfections is one of the advantages of glulam over lumber.

Thick lumber is relatively expensive to dry because drying time increases out of proportion with increasing thickness. Many builders are unwilling to pay for drying, installing thick lumber green instead to dry in use under uncontrolled conditions, which causes checks and involves the risk of decay. Furthermore, green lumber must be designed for correspondingly low allowable stresses. By contrast, glulam consists of lumber that was dried defect-free in thin laminations before assembly; it has high allowable stresses and small cross sections.

The moisture content at the time of gluing must be below 16%; the upper limit is 20% for timbers to be used at equilibrium moisture contents above 16%. In use the moisture contents inevitably change, causing the laminations to swell and shrink differently and to induce internal, strength-reducing stresses. To keep such stresses small, all laminations should be of the same species and grain, free of abnormally shrinkage-prone reaction wood, and of the same moisture content at gluing time, not differing more than 5%. These restrictions add to the cost of the product and so are not strictly observed for interior use, where moisture contents vary little.

Preservation, Gluing, Quality Control, and Applications

Glulams that remain wet for extended periods or are exposed to the weather have to be impregnated with a wood preservative. The industry treats the boards before assembly with a preservative that is compatible with the glue. Treatment after gluing is difficult because the preservative has a long way to go to reach center, flowing in as it does parallel to the penetration-retarding gluelines.

As for water resistance of the glue bond, the standard differentiates between *dry-use* adhesives for glulam whose moisture content is not expected to exceed 16%, and *wet-use* adhesives otherwise. The standard restricts the types of adhesives and gives rather specific instructions for the laminating process.

Manufacturers have to maintain an elaborate quality control system with in-line tests, physical tests, visual inspection, and certification by a qualified inspection agency. This adds to the cost of the product but the resulting reliability is worth some expense.

Glulams find many more applications and are engineered farther than build-up beams; their production is expected to increase substantially in future. In their large dimensions and beauty, laminated timbers are very well suited for structures with wide spans, such as roofs of athletic competition buildings, assembly halls, gymnasiums, warehouses, factory halls. Architects like glulam for churches and all kinds of dome-shaped structures. In each of these cases *horizontal* laminations (Figs. 5.2, 5.3, and 5.4) are most practical. Bridges with small spans, which can be designed without supporting arches, are best handled with straight *vertical* laminations, as illustrated in Fig. 5.3, since the whole deck then consists of one compact piece.

Figure 5.4. Dome of Northern Arizona University's stadium *Ensphere* in Flagstaff, with a clear span of 502 ft, the world's largest glulam dome. Photo shows the dome before installation of the center skylight framing. (Photograph courtesy of Western Wood Structures, Inc.)

In the United States roughly 30 firms make structural glued laminated timbers. They are inspected by their association, the American Institute of Timber Construction (AITC, Englewood, Colorado), whose quality inspection mark is well-known in timber construction.

5.4. VENEER

Veneer is a thin sheet of wood. Most sheets serve as the 1/16 to 3/16 in. thick plies of plywood. Very valuable sheets, cut 1/32 in. or thinner as decorative *face* veneer, are bonded to the surface of plywood or of other core materials to improve the surface. For special purposes the industry cuts wood as thin as 5/1000 in., but such thin sheets tend to break and call for special handling equipment, or must be paper-backed or paper-overlaid, as in veneer wallpaper and veneer calling cards. Wirebound boxes and bushel baskets consist of single veneer layers up to 1/4 in. thick.

Most veneer is *peeled* (*rotary-cut*) on a lathe that rotates the log against a razor-sharp knife, cutting a continuous sheet just like unwinding a roll of paper (Fig. 5.5*a*). The knife cuts approximately along annual rings, so that the rings appear far apart and form a a rather obtrusive *figured* pattern in species with

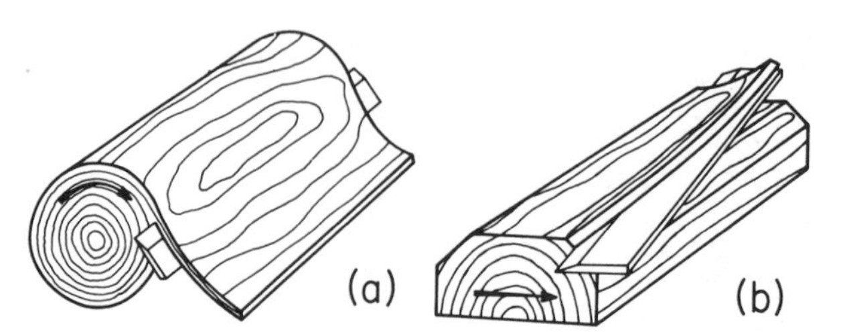

Fig. 5.5. (*a*) Veneer rotary-cutting. (*b*) Veneer slicing.

distinct growth rings, such as southern yellow pine and Douglas fir. The plywood industry prefers rotary-cutting because of its low cost.

Veneer mills *slice* valuable logs into face veneer. The slicing machine cuts individual sheets by up and down or back and forth movement of wood against the knife, or by movement of the knife against wood (Fig. 5.5*b*). In sliced veneer the annual rings lie close together and create an attractive *plain* pattern in many wood species. *Quarter slicing* attempts exactly radial cuts to expose ray bands, which add special configurations to oak and other species. In interlocked grain timber (Section 1.6), quarter slicing brings out ribbons of varying fiber deviation, which reflect light differently and contribute to the high value of many tropical woods, including mahogany. The mills assemble the veneer sheets as *flitches*, in the same order they came from the log. The term *flitch* also designates the log section made ready for slicing by shaping up its edges (Fig. 5.5*b*).

The knife bends the veneer aside and tends to break it along longitudinal *knife checks*, particularly in rotary-cutting where the weaker planes of rays run perpendicular to the cutting plane (Fig. 5.6). A pressure bar (*P* in the Figure) or a pressure roll compresses the wood to reduce checking and to minimize splitting ahead of the knife edge. Heating the logs in steam or in hot water has the same beneficial effects and also reduces cutting resistance. Light, uniformly structured species check less and cut more easily than dense kinds of wood with high earlywood-latewood contrast, such as southern yellow pine and ring-porous hardwood. Good thin face veneer is free of knife checks (*firm-cut, tight*), but knife checks must be expected in sheets thicker than 1/32 in. Test for checks by bending alternately to both sides; checks facilitate bending away from the loose toward the tight surface.

Sawed veneer would be tight regardless of thickness, but the surface roughness and particularly the loss of wood in sawdust do not justify the gain; sawed veneer disappeared from the market mainly for this reason. Knife checks are an aesthetic drawback in face veneer, and they extend deeper when the glue-bonded veneer shrinks under restraint. Since rotary-cut veneer shrinks in width by the larger tangential magnitude, its knife checks increase in service much more than do those of sliced veneer.

Commercial consumers of veneer can find suppliers in a brochure of the Hardwood Plywood Manufacturers Association (P. O. Box 6246, Arlington, Va

Figure 5.6. Detail of veneer-cutting operation on a rotary lathe, much enlarged. P, pressure bar; K, knife. (Photograph courtesy of U.S. Forest Products Laboratory.)

22206). The brochure is titled *Where to Buy Veneer* and lists the manufacturer along with the wood species, as well as the various thicknesses and sizes. Hobbyists who need only a few sheets buy from special suppliers, e.g., Craftsman Wood Service Co., 1735 W. Cortland Ct., Addison, Ill. 60101, or Albert Constantine and Son, Inc., 2050 Eastchester Road, Bronx, N.Y., N. Y. 10461. Lumber dealers and do-it-yourself stores rarely ever carry veneer.

6

Wood-Base Panels

In 1980 the volume of lumber consumed in the United States was three times larger than the volume of plywood, particleboard, and fiberboard together. Use of the three panel products is rising however while that of lumber stagnates. In early times man used timber for construction mainly as it grew, in the round, or squared out of the round with an ax. During the industrialization period sawmills converted increasing amounts of timber into lumber. Later in this century the industry modified the raw material farther and developed wood-base panels to meet specific needs and to use timber more efficiently. Plywood appeared first, then fiberboard, and finally particleboard after World War II. Plywood tops the others in significance and will be treated here first, before proceeding to particleboard and fiberboard, following the sequence of increasing disintegration of timber in the manufacturing process. Finally the panel products are compared with lumber.

6.1. PROPERTIES OF PLYWOOD

Plywood is a crossbanded assembly of layers joined with an adhesive. In *veneer plywood* the layers consist of veneer—three of them in the most simple and most common, 1/8 in. to 3/8 in. thick panel (Fig. 6.1). In other plywood, veneer is combined with a core of lumber (*lumber core plywood*, Fig. 6.2), particleboard, hardboard, or any other suitable material. The arrangement in alternating fiber directions (*crossbanding*) represents plywood's essential feature.

Strength and Stiffness

Plywood is as strong and stiff as its layers combined. Consider the panel of Fig. 6.1, in which the fibers of *face* and *back* are oriented in direction A, while the *core* fibers point in direction B. Consequently plywood strength in direction A is inbetween that of lumber in fiber direction and that of lumber perpendic-

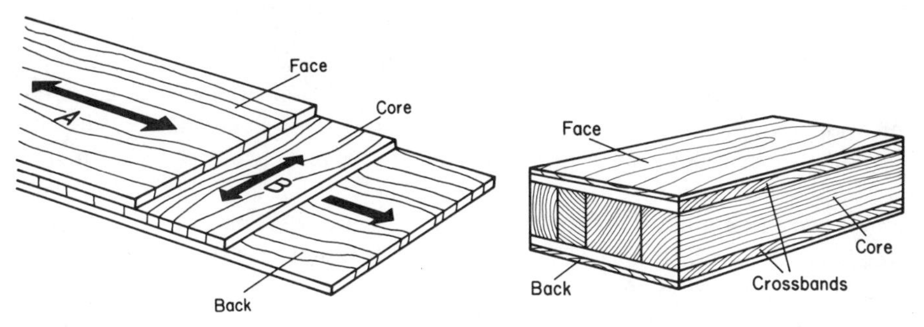

Figure 6.1. Orientation of veneer in three-ply plywood.

Figure 6.2. Five-ply lumber core plywood.

ular to the grain. In other words, plywood strength exceeds lumber strength perpendicular to the fibers but does not reach the longitudinal strength of lumber. Exact strength in directions A and B depends on the thickness of the plies. In common 3-ply panels, whose two surface veneers are together as thick as the core, tension strengths in the two directions equal the average strength of the veneers in fiber direction and perpendicular to fibers. Since strength perpendicular to the fibers is negligibly small, particularly in knife-checked veneer, the panel has half the strength of solid wood in fiber direction. Bending strength in the surface fiber direction (direction of panel length) exceeds half because, in bending, surface veneers carry the brunt of the load. The bending properties in direction of panel width are correspondingly lower.

Having nonetheless an adequate strength in width direction, plywood can be thinner than lumber for many uses. Consider, for example, subflooring under the impact of a youngster jumping: using lumber a single narrow board may have to absorb most of the impact, whereas using plywood it is usually a panel 4 ft wide. Plywood sheathing 3/8 in. thick will serve the purpose as well as 1 in. lumber. The crossbanded panel can be thinner also because it does not split: splits want to spread in fiber direction, not perpendicular to the grain as they have to in the transverse layers. Freedom from splitting permits nailing close to edges; in boxes made of 1/4 in. plywood, nail joints have the strength they would have in 3/4 in. lumber. Transverse to the plane (in direction of thickness) plywood strength naturally does not exceed that of lumber.

Dimensional Stability

Crossbanding drastically alters shrinking and swelling in the plane. Again see Fig. 6.1 and consider direction A. Taken by themselves, the face and back will shrink very little, since it is their longitudinal direction, while the transverse core shrinks many times more. It is the glue bond that forces all three veneers to shrink by one and the same amount: the face and the back restrain core shrinkage, while the core in turn stretches both surface veneers. As a result, the panel shrinks to a degree somewhere between that of solid wood in fiber direction and solid wood across the grain. However, since in direction A the two surface veneers are stiff and resist compression by the weak core, the plywood actually shrinks very little, hardly more than solid wood in fiber direction. It resembles a sandwich of stretched rubber between two steel plates. According to rule of thumb, crossbanded panels shrink and swell in the plane about two times more than lumber in fiber direction. When installing dry plywood sheathing which may be expected to adsorb moisture, leave 1/16 in. space at all edge and end joints to prevent buckling.

In its thickness, each layer of plywood shrinks freely by the same percentage as lumber, unrestricted by adjoining layers. Hence the shrinkage of plywood assembled from rotary-cut veneer should equal the radial shrinkage of lumber. Hot-press compression however may *spring back* in its first swelling-shrinking cycles, with the effect that new plywood tends to swell more and to shrink less than lumber.

Dimensional stability in its own plane may mislead one to think that plywood checks less than lumber. A surface checks when inner layers keep it from shrinking. The inner layers of lumber exert restraint when they dry more slowly than the surface, whereas in plywood restraint results from the crossbanding, no matter how fast the surface dries. Hence plywood may check under mild conditions that would cause no checks in lumber.

Checking relieves the tension stress that caused it. Relief in the thin layers of plywood, unlike relief in lumber, is restricted to the exact area of the check; the neighborhood remains restrained and so checks too. The number of checks in plywood is thus much higher than in lumber.

By contrast however to checks in lumber, those in plywood are restricted to face veneer only, not extending into the layer below; and the thinner the veneer, the shallower the checks. Most drying checks develop from knife checks which occur more often in thick veneer, so that by comparison drying checks in thin face veneer are not only shallow but also rare. How much plywood and veneered panels will check depends also on the location of knife checks. Any tiny check at the very surface acts as a notch and deepens when the panel dries. Sanded panels of high grade veneer have the tight surfaces out.

Surface checks are a serious problem in cabinets and furniture and other products where a smooth finish film only accentuates the defect. Fresh liquid finish covers checks, but later, when the moisture content of the face veneer drops, these checks open, extending into the film and cracking it, since no hard finish is flexible enough to bridge the gaps. Plywood should therefore be very dry at the time of finishing. The film will generally be flexible enough to stretch when the panel later adsorbs moisture and swells a little. In disputes arising over cracked finish films, finish suppliers tend to blame the plywood, while plywood suppliers accuse the finish of being too brittle. Film cracks located exactly above a veneer crack can be presumed to result from the check, whereas when cracks occur above intact veneer, the finish has lacked enough flexibility to adapt to a small plywood swelling.

Warp

Plywood warps due to one-sided swelling and shrinking, the one-sidedness having two reasons: (1) one-sided moisture changes, either from exposure on one side only or because moisture contents of face and back were different at the time of bonding; (2) one-sided swelling and shrinking in spite of identical moisture conditions on both sides—due to cross grain, different species, different thickness, or natural variability in shrinkage accentuated by relatively thick veneer. Among these factors, cross grain causes *twist* (one corner rising out of the plane formed by the other), while the other three factors account for *bow* (the panel's central area rising out of the plane formed by the four corners).

6.2. PLYWOOD VARIETIES

The industry makes plywood of many wood species and species combinations, from veneer of different thickness and quality, perhaps with cores other than veneer, using different glue bonds, with additives for improved durability, then sanding, finishing, or overlaying the surface. Variables such as these are discussed in this section.

Thickness and Number of Layers

Thick plywood can be made very cheaply from three relatively thick layers, but such panels tend to warp and check, as explained in the preceding section. Besides which, under changing moisture conditions these thick layers exert strong shear force on the glue bond and impair its durability. Furthermore the panel grade mainly depends on the grade of surface veneer, which the industry cuts from rare high-grade veneer logs. Adjoining inner layers cannot be much thicker than the surface veneers because the latter must restrain the core's transverse swelling. For these reasons, thick plywood must consist of more than three layers (Fig. 6.3). To *balance* the panel and avoid warp, the number of layers should be same at each side of the core layer. Plywood is therefore composed of an odd number of layers.

To simplify production some mills use layers of two or more *plies* lying in the same orientation, cutting only one thickness of veneer and layering to reach different thicknesses in 3/8, 1/2, 5/8, 3/4, and 7/8 in. plywood, all using 1/8 in. veneer for example. The 1/2 in. panel consists of four plies, two of which form the parallel laminated core layer, while the 5/8 in. kind is made of five alternating plies. The 3/4 in. panel features six plies, again two in the core

Figure 6.3. Five-ply veneer plywood.

oriented in parallel. Some 7/8 in. plywood has outer layers from two parallel laminated veneers—face and *sub-face* forming one layer, back and *sub-back* the other surface layer—to gain high strength and bending stiffness in fiber direction of the surface veneers, as needed for subflooring. The number of plies can be even, but the number of layers has to be odd for the sake of balance.

Many plies consist of two or more butt-jointed (spliced) veneer strips. The mills edge-bond the strips with glue, tape, or thread to make handling easier and to prevent them from shifting out of place before pressing. The edge joints of a panel should be tight but joint strength does not matter, since adjacent layers hold the strips. In panels of high appearance grade, the number of strips is limited and the strips must match for color and grain.

Standards and Grades

Most plywood produced in the United States is graded according to the two *Product Standards* (PS) *1-74* (*Construction and Industrial Plywood*) and *51-71* (*Hardwood and Decorative Plywood*). Both are voluntary but generally accepted; they differ according to different uses: *PS 1-74* plywood (formerly *Softwood Plywood*) for construction, and *PS 51-71* plywood where appearance is crucial, for example in panelling and furniture (Table 6.1). *PS 1-74* and *PS 51-71* overlap; the *Construction* plywood standard includes decorative panels with striated, grooved, embossed, and brushed surfaces, while some *Hardwood* plywood serves structural purposes too, in trailers and other places.

Many *Construction* grade designations include the grades of the surface veneers. In the example *A-D* the face is of grade *A*, the back of the low grade *D* as permitted for built-ins where it makes no difference whether the back has large knots and other defects. Grades of inner plies fall below the grade or equal the grade of the back veneer in practically all grades of *Construction* plywood. The highest veneer grade is *N*, used as the face of plywood intended for natural finish.

The *Hardwood* plywood standard defines no panel grades except veneer grades: *Premium* (*A*), *Good* (*1*), *Sound* (*2*), *Utility* (*3*), and *Backing* (*4*); these are little known. Many large consumers such as furniture manufacturers often write their own specifications and disregard the grades.

Water Resistance Types

When exposed to moist conditions, some plywood fails in the glueline under low load. The layers may even separate (delaminate) without any external load whatever. The standards recognize this fact and use almost identical terms for the various degrees of panel integrity, defining an *Interior* and an *Exterior* type. Panels that fail to meet even the *Interior* requirements may be suitable for some purposes but are not an article of commerce. The *Interior* type is intended as the name implies for interior application; the *Construction* plywood standard subdivides it into three levels of durability:

1. Bonded with *interior* glue, corresponding to the *Interior* type of *Hardwood* plywood.
2. Bonded with *intermediate* glue "intended for protected construction . . . where moderate delays in providing protection may be expected."
3. Bonded with *exterior* (waterproof) glue for plywood "intended for protected construction where protection against exposure due to long construction delays . . . is required."

The *PS 1-74* waterproof *Exterior* type "will retain its glue bond when repeatedly wetted and dried or otherwise subjected to the weather and is, therefore, intended for permanent exterior exposure."

The effect of water mainly depends on the kind of glue. Urea resin gives the resistance of the *Interior* type, whereas *Exterior* plywood requires the more expensive *waterproof* phenolic resin. These are the most common glues, but the mills may bond with any other kind or combination of glues. The plywood type is not defined by its glue, rather by performance of samples in standardized soak-and-boil tests. Since phenolic resin is reddish brown, at face value a glueline of this color speaks for a waterproof bond, but the factory may have mixed phenolic resin with large amounts of a cheap extender, or the glue may have cured before pressing, or the layers were too thick, or the factory made some other gluing mistake. Plywood that delaminates under exposure to the elements is not *Exterior*. *Interior* plywood bonded with *exterior* glue does not meet the *Exterior* requirements because the veneer grades are too low for strong swelling and shrinking.

Table 6.1. Comparison of the Plywood Standards PS 1-74 and PS 51-71

Designation	PS 1-74 Construction and Industrial Plywood	PS 51-71 Hardwood and Decorative Plywood
Application	Construction and industrial, including decorative panels with striations, grooving, etc. of the face veneer	Decorative wall panels, furniture, cabinets, containers, etc.
Species classification	Stiffness Groups 1 through 5 formed to facilitate balanced construction	Categories A, B, and C formed to identify maximum veneer thickness
Face species	Softwoods and hardwoods	Primarily hardwoods; decorative veneers of various softwoods
Back species	Same Group as face	Any hardwood or softwood
Inner plies	Veneers of Group 1 through 5 or any species with specific gravity of 0.41 or more	Veneer or lumber core of any species, particleboard, hardboard, or other proper material
Main species	Douglas fir and southern pine	Many
Veneer grades	N for natural finish A smooth and paintable B solid surface C knotholes up to 1 in. C PLGD C with plugs D knotholes up to 2-1/2 in.	A Premium 1 Good 2 Sound 3 Utility 4 Backing SP Specialty, e.g., wormy chestnut

Plywood type and grade	Interior N-N ... C-D Underlayment Structural I and II Exterior Marine A-A ... MDO Special Exterior A-A ... MDO A-A ... C-C A-A HDO ... Special Overlays	Technical - (Exterior) Type I - (Exterior) Type II - (Interior) Type III - (Interior)
Designation example	Interior A-D according to grades of face and back veneer	Walnut (1) Type II according to species and grade of the face veneer
Thickness	1/4 in. to 5/4 in.	1/8 in. to 3/4 in.
Thickness of veneers	Minimum from 1/16 in. to 1/10 in. depending on panel thickness; face and back not thicker than 1/4 in.; inner layers not thicker than 5/16 in. Minimum number of plies in some Exterior grades, e.g., 7 plies for thickness over 3/4 in.	Maximum veneer thickness depending on species category, up to 1/4 in.; dense species less thick

Lumber Core and Crossbands

Many table tops and thick panels of cabinets consist of *lumber core plywood* (Fig. 6.2), a kind included in the *Hardwood* plywood standard. The core strips cost less than veneer of same thicknesses, but they may swell and shrink in the panel's thickness by different amounts. In addition, wide flat-grained strips will cup. These different swellings and cuppings tend to *telegraph* through the face veneer, so that stripes of uneven light-reflection reveal the core structure beneath. *PS 51–71* requires therefore that all strips in a core be of the same wood species and not more than 2.5 in., at most 4 in. wide, depending on the species category.

Cores located immediately under thin face veneer practically always telegraph. Additionally, the large transverse swelling of thick cores breaks the thin face, while shrinking cores of this kind cause the face to crumple in compression, unless a relatively thick *crossbanding* veneer is sandwiched between core and face (Fig. 6.2). The term *crossband* is used also for layers under the surface of other plywood, particularly of panels with five or more layers. In 3-ply plywood, *crossband* means simply *core.* Some people use *crossbands* to refer to all the layers whose fibers run perpendicular to those of the outer plies.

Spans, Allowable Stresses, and Industry Associations

Subflooring and roof sheathing panels carry loads that deflect them, depending on the span of the supporting frame. *PS 1–74* lists maximum spans in an *Identification Index* table which tells, for example, that for 1/2 in. grade *C–D* panels in the wood species groups 2, 3, and 4, the roof frame spacing can be at most 24 in., while the same panel is too weak for subflooring. Spans (*24/0* in Fig. 6.4) are frequently included in the grade stamp.

Exterior and *Structural Interior* grades of *Construction* plywood are suited for applications where tension, compression, shear, and nail bearing have significance. The corresponding allowable stresses are listed in brochures of the *American Plywood Association* (*APA*), which also include live loads for roof decking, stud spacing in wall sheathing, and numerous general engineering data. For the convenience of architects, engineers, and builders, the *APA* has compiled design requirements of major building codes and the *Minimum Property Standards* of the *U.S. Department of Housing and Urban Development* (*HUD*).

The *APA*, P.O. Box 11700, Tacoma, Wash., 98411, affiliates most producers of *Construction* plywood while the *Hardwood* plywood industry is organized in the *Hardwood Plywood Manufacturers Association* (*hpma*), P.O. Box 2789, Reston, Va., 22206. Both associations assist users of plywood with literature far beyond the mentioned engineering data, including examples of plywood uses with detailed plans, design outlines, use guides, hints for cutting expenses, specifications, instructions for finishing, lists of suppliers of finishes, descriptions of building components and complete structures, grade guides

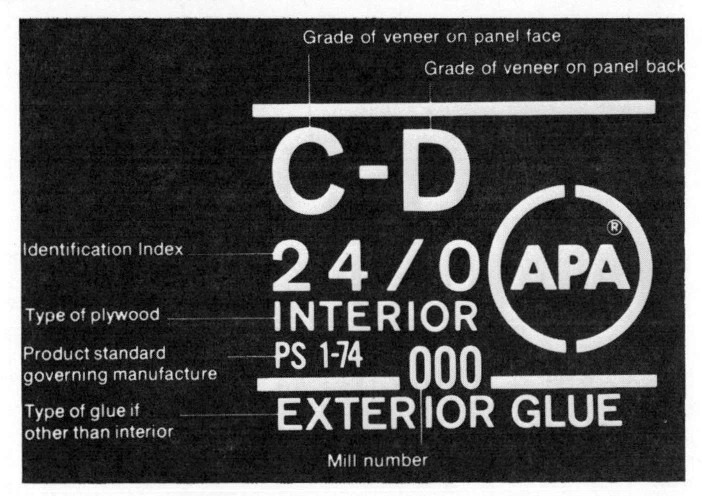

Figure 6.4. *Construction and Industrial Plywood* grade stamp. (Photograph courtesy of American Plywood Association.)

(Table 6.2), descriptions of special products such as curved, fire-retardant-treated, fiberglass-reinforced-plastic, overlaid, and preservative-treated plywood. The *hpma* brochure *Where to Buy Hardwood Plywood & Veneer* lists sizes, types, wood species, plywood specialties, and so on.

Both associations test panels and give manufacturers the right to use the associations' trademarks *APA* (Fig. 6.4) and *hpma*, respectively. Association grades comply with the standards, except that members of the *APA* offer additional proprietary grades. The *APA* and the *hpma* are not the only organizations which test and supervise the quality of plywood. Some producers have their panels controlled by special testing firms such as *Teco*. The trademarks appear in the grade stamp on the back or along the edge of each panel.

6.3. PARTICLEBOARD

Plywood, particleboard, and fiberboard consist of bonded pieces: plywood of layers as large as the panel, fiberboard of fibers too small to be seen with the unaided eye, and particleboard of discrete particles, which run typically about ½ in. large, with some up to 10 in. long and others almost invisibly small.

Particleboard is sometimes called *chipboard, chipcore, flakeboard, shavingsboard, waferboard,* and *splinterboard,* depending on the particles. Other occasional names include *composition lumber, synthetic lumber,* and *coreboard.* In the early years of particleboard history, people could not agree about a generic name until the standard *CS 236-66* (*Mat-formed Wood Particleboard*) settled on *particleboard,* and recommended terms like *flake-type particleboard* for more specific designations based on the contents.

Table 6.2. Grade-Use Guide for Interior-Type *Construction and Industrial Plywood* (Photograph courtesy of American Plywood Association)

Grade Designation	Description and Most Common Uses	Typical Grade-trademarks	Veneer Grade			Most Common Thicknesses (inch)					
			Face	Inner Plies	Back						
N-N, N-A N-B INT-APA	Cabinet quality. For natural finish furniture, cabinet doors, built-ins, etc. Special order items.	N-N G-1 INT-APA PS 1-74 000 N-A G-2 INT-APA PS 1-74 000	N	C	N,A, or B						3/4
N-D-INT-APA	For natural finish paneling. Special order item.	N-D G-2 INT-APA PS 1-74 000	N	D	D	1/4					
A-A INT-APA	For applications with both sides on view, built-ins, cabinets, furniture, partitions. Smooth face; suitable for painting.	A-A G-1 INT-APA PS 1-74 000	A	D	A	1/4		3/8	1/2	5/8	3/4
A-B INT-APA	Use where appearance of one side is less important but where two solid surfaces are necessary.	A-B G-1 INT-APA PS 1-74 000	A	D	B	1/4		3/8	1/2	5/8	3/4
A-D INT-APA	Use where appearance of only one side is important. Paneling, built-ins, shelving, partitions, flow racks.	A-D GROUP 1 APA INTERIOR PS 1-74 000	A	D	D	1/4		3/8	1/2	5/8	3/4
B-B INT-APA	Utility panel with two solid sides. Permits circular plugs.	B-B G-2 INT-APA PS 1-74 000	B	D	B	1/4		3/8	1/2	5/8	3/4
B-D INT-APA	Utility panel with one solid side. Good for backing, sides of built-ins, industry shelving, slip sheets, separator boards, bins.	B-D GROUP 2 APA INTERIOR PS 1-74 000	B	D	D	1/4		3/8	1/2	5/8	3/4
DECORATIVE PANELS—APA	Rough-sawn, brushed, grooved, or striated faces. For paneling, interior accent walls, built-ins, counter facing, displays, exhibits.	DECORATIVE B-D G1 INT-APA PS 1-74	C or btr.	D	D		5/16	3/8	1/2	5/8	
PLYRON INT-APA	Hardboard face on both sides. For counter tops, shelving, cabinet doors, flooring. Faces tempered, untempered, smooth, or screened.	PLYRON INT-APA 000		C & D					1/2	5/8	3/4

Numerous molded products such as toilet seats and cricket balls consist of bonded wood particles and are *particleboard* in a broad sense, their features resemble those of the flat panels to which this section is restricted.

Manufacture, Kinds, and Structure

You can make particleboard yourself: spray glue on planer shavings, form the blend into a mat, compress the mat between platens until the glue has cured, and trim the edges. The first particleboards were made in this manner around World War II in timber-short Europe, but since then the industry steadily has improved the manufacturing process as well as the panels.

The particles were improved first. In curly planer shavings hardly any binder could reach the inner surface, also the pieces overlapped too little for adequate bonding. The shavings had to be crushed by hammer milling before blending. Slender flakes cut in fiber direction from sawmill edgings, slabs, and roundwood gave still stronger panels, while of all major wood residues, sawdust turned out to be least suitable.

Since particles near the surface contributed much more to bending strength and stiffness than core particles, the industry departed from *homogeneous* boards to make *3-layer* panels with engineered flakes and slivers in the surface and cheap coarse particles in the core. Progressive plants, ever trying to gain an advantage over competitors, composed *5-layer* panels and finally *graduated* panels in which the particles gradually changed in size and shape from the surface to the center.

Large particles in the surface make the board stiff and strong, but they telegraph through overlays and through thin face veneer. *Corestock* panels for veneering need a smooth layer of *fines* (e.g., wood flour) at the very surface to prevent telegraphing. Smooth surfaces are essential also for imprints of wood grain and of other patterns (Section 14.9). Surface characteristics rather than strength and stiffness in fact determine the quality of many boards.

Particleboard properties depend on density, as do the properties of lumber and plywood. In turn, board density depends on mat compression in the hot-press. Most manufacturers aim at a density of 37 lb/ft^3, which for this reason has been chosen as the borderline between *medium* and *low density* board (Table 6.3).

The glue (*binder, resin*) does not really coat the blended particles, instead small droplets—visible at the surface of some panels—cover fractions of the particle surface. Complete coating would provide stronger, more water resistant bonds but it would require more glue than wood, and the *wood-filled resin* would be much more expensive and heavier than particleboard (which contains 3 to 9% solid resin, *solid* meaning *without solvent*). Even with the average content of 6% the resin costs as much as the wood, or one third of all manufacturing expenses.

Most particleboard is mat-formed (flat-platen pressed) as described above (Fig. 6.5). Americans make less than 1% in the other, cheap extrusion process,

Table 6.3. Property Requirements[a] for Mat-Formed Particleboard.

Type	Density lb/ft^3	Class	Modulus of Rupture (psi)	Modulus of Elasticity (psi)	Internal Bond (psi)	Swelling[b] (%)
1[c]	≥50 (A)	1	2400	350,000	200	0.55
		2	3400	350,000	140	0.55
	37-50 (B)	1	1600	250,000	70	0.35
		2	2400	400,000	60	0.30
	≤37 (C)	1	800	150,000	20	0.30
		2	1400	250,000	30	0.30
2[d]	≥50 (A)	1	2400	350,000	125	0.55
		2	3400	500,000	400	0.55
	37-50 (B)	1	1800	250,000	65	0.35
		2	2500	450,000	60	0.25

[a]Averages per panel according to CS 236-66; minimum strengths and moduli of elasticity.
[b]Maximum swelling due to relative humidity increase from 50 to 90%.
[c]Generally made with urea-formaldehyde resin binders, suitable for interior applications.
[d]Made with durable and highly moisture and heat resistant binders, generally phenolic resin.

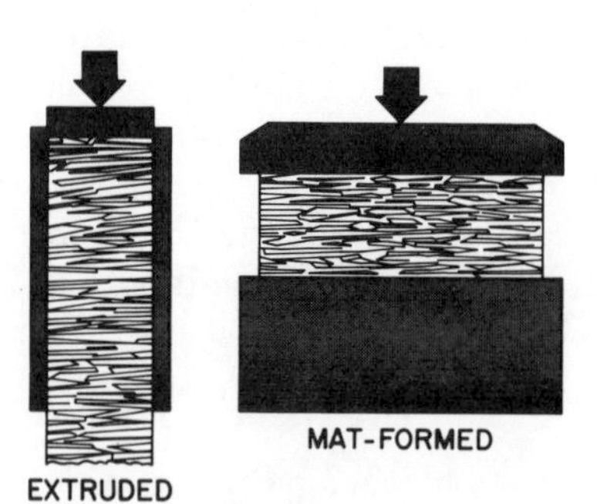

Figure 6.5. Manufacture of mat-formed and of extruded particle-board. (Photograph courtesy of J. D. McNatt.)

which resembles squeezing toothpaste out of the tube: a bar moving up-and-down pushes the blended particles little by little through a long wide hot die (Fig. 6.5), forming an endless board as thick as the die's narrow gap between the two large plates.

During assembly, elongated and flat particles fall horizontally in both processes (Fig. 6.5) like toothpicks and cards on the floor. There is however a difference in the finished panel of each process: in the mat-formed kind the particles lie within the panel plane like veneers in plywood, whereas in the extrusion process some elongated particles point in panel thickness, others have panel width orientation, and many fall between these two directions, but few lie lengthwise. The orientation determines panel quality, as we shall see.

Strength and Stiffness

Spotty glue bonds provide relatively little strength. The boards are especially weak in tension perpendicular to their long and flat particles, as measured in *internal bond* tests. Tensile strength perpendicular to the plane as such does not matter, since particleboard panels are not stressed this way in use. However, in the internal bond test the specimen fails in the weak core, on whose cohesion depends how well the panel edges can be machined and hold screws, of cabinet door hinges for example. Many edges must be reinforced with an edge strip or covered with an edge band. Users check internal bond simply by forcing a knife blade into the edge and twisting the blade, trying to split the board in its center. Screws in the panel face (*face screws*) are held by and depend on the strength of all layers, not only that of the weakest layer, as with internal bond. Face-screw withdrawal resistance is still inadequate in many cases.

In extruded board the particles oriented to panel thickness provide high internal bond but little strength in the plane, and require veneering for adequate flexural properties. Furniture factories use extruded board as *corestock* (center ply, to be veneered), making it on the side from their own scraps by the relatively cheap extrusion process equipment.

Wafer-type particleboard, known as *waferboard*, deserves special consideration. The wafers are several inches long and nearly as wide, resembling veneer and supposedly giving a panel as strong as plywood. Like particles in other mat-formed board the wafers lie oriented at random in the panel plane and provide strength mainly in that plane, as needed for house sheathing.

Oriented structural board (OSB) consists of long oriented particles, aligned by means of an electrical field, or aligned mechanically by vibrating the particles through fins. OSB in which all particles lie in direction of panel length resembles solid wood; it excells in lengthwise strength at the expense of widthwise strength, and is intended as a load-bearing member in construction. Other OSB has two surface layers of lengthwise particle orientation and a core with particles oriented in the panel width direction acting as a crossband for high lengthwise bending strength. The similar commercial panel *plystran* features two veneers at the surface and an OSB crossband core.

Moisture weakens particleboard at least as much as it weakens solid wood. Whether the effect is temporary and reversed by redrying or permanent depends on the glue bond, as explained in the following paragraphs.

Dimensional Stability and Moisture Resistance

Common board made of long or flat particles swells in length and in width about 0.6% over the whole moisture range, little more than plywood but much less than lumber across the grain; it is crossbanded and fairly stable dimensionally. In form stability (resistance to warp) good particleboard may surpass plywood, whose symmetry is disturbed by natural variations in the raw material. OSB consisting of layers with alternating orientation naturally swells and shrinks like ordinary particleboard, while uniform OSB moves like solid wood.

Panel thickness stability is an entirely different matter. Each long particle changes in thickness more than lumber does transversely, because in addition to normal swelling it springs back from the compression of the hot-press—later—that is, during moisture gains and in swelling-shrinking cycles. Since most panels are hot-pressed into 20 to 30% greater density than the wood species they come from, the spring-back can be substantial. To decrease water absorption and the swelling rate, manufacturers add about 1% wax (based on particle weight) to the binder. To some extent the wax does also reduce adsorption of vapor; on the other hand, it tends to interfere with the binder and impairs the strength of wet panels.

Particleboard manufacturers can increase strength by compressing the mat only to the point where spring-back and thickness swelling become excessive. This is unfortunate, since compressing strengthens the panel in an efficient way, through interlocking and entangled overlapping of particles.

Extruded board swells little in thickness but much in the plane; it warps considerably more than common particleboard, because its swelling varies more between opposite surfaces and between spots in the same surface.

As in plywood, the moisture resistance of particleboard depends mainly on the binder used. *CS 236-66* distinguishes between panels of *type 1*, generally made with urea-formaldehyde resin binder for interior application, and *type 2* panels made with durable and highly moisture- and heat-resistance binders, generally phenolic resin, for interior and certain exterior applications. As a result of incomplete coating of the particles, particleboard is not quite as moisture resistant as plywood.

Sizes, Uses, and Grades

Dealers offer the panels in thicknesses from 1/8 to 2 in., most are 1/4 to 3/4 in. thick, as needed for furniture corestock, floor decking in mobile homes, and floor underlayment. The general size is 4 × 8 ft. For industrial use the manufacturers cut many panels to specified sizes.

Particleboard types and grades (Table 6.3) are hardly known on the retail market. Commercial consumers use the grades typically as framework and add their specific needs through additional upgrading requirements, specifying higher strengths and special features such as *fire-resistance* and *durability*, which the producers then achieve by adding fire-retardants and preservatives to the binder.

Among the grades in Table 6.3, *type 1 density B class 1* (*1B1*) represents essentially floor underlayment; it is not strong enough for corestock, the typical use of grade *1B2*. The weakest grade (*1C1*) serves as a core for doors which gain their bending strength through veneer skins and the lumber frame. The strongest grade (*2A2*) is needed for corestock which has to resist very high pressure when being overlaid with *high pressure laminates.* Grade *2B1* was developed as a basically exterior panel for siding, with pressed grooves and high-resin fiber overlays, the edges machined for suitable joints.

The *National Particleboard Association* (2306 Perkins Place, Silver Spring, Md., 20910) grants members the right to use the *NPA* trademark, which guarantees the panel meets the requirements of the designated *CS 236–66* grade. The association supervises member mills to make sure that customers receive what they pay for.

6.4. FIBERBOARD IN GENERAL

Fiberboard is a group of wood-base panel products which greatly differ in appearance, mechanical properties, and uses. This section concerns the common features of all fiberboard, with two categories discussed later, *insulating board* and *hardboard*.

Manufacture and Appearance

Fiberboard properties are best understood through an understanding of the paper making process in which matchbox-size wood chips are disintegrated into fibers (pulped) by water-dissolved chemicals penetrating the chips in pressure cookers to dissolve lignin, the substance that bonds fibers together. The delignified separated fibers are washed and suspended in large amounts of water; the resulting slurry flows onto a fine wire-mesh screen through which most of the water drains, leaving the fiber pulp to form a thin wet mat. Peeled from the screen and dried, this thin mat is nothing less than paper, whose fibers hold together by interfelting and by natural bonds (particularly hydrogen bonds) without any added binder.

Fiberboard factories fiberize chips and roundwood bolts mechanically, without lignin-dissolving chemicals, leaving the lignin in. Hot water or steam softens the wood before the attrition mill, a sort of grinding machine, rubs the fibers apart. Defibering is less perfect than pulping with chemicals; fiberboard pulp contains fiber fractions (*fines*), bundles of fibers, and larger fiber aggregates which are not tolerable in pulp meant for good paper.

The pulp has then to be washed to remove wood extractives along with other solubles and fines, leaving the fibers suspended in a water slurry, which runs on a wire mesh screen to drain off water as in paper manufacture. Due to the imperfect defibering and the stiffening effect of lignin the fibers interfelt and bond less than paper fibers. The board still has sufficient strength for many uses; but some kinds need to be enforced by adding 2 or 3% binder to the slurry. Finally the water-drained mat is compressed and dried in various ways, then edged to make square rectangular panels.

The fiber aggregates, varying in size and number among the particular processes and board categories, determine the appearance of the board surface. In insulating board where the aggregates are almost too small to be seen with the unaided eye, the board looks like an homogeneous compacted, fibrous mass. In hardboard, the dense category of fiberboard, fiber aggregates appear as pieces of various sizes imbedded in masses of indistinct fibers; some aggregates have the shape of needles and may be half an inch long, while others look like tiny print dots, with a wide variety of sizes and oddly-shaped forms inbetween.

Dimensional Stability

The mat-forming fibers land horizontally on the screen, lying in the plane of the mat; only a few point in the direction of mat thickness. Fiberboard is crossbanded like plywood and mat-formed particleboard; it shrinks and swells in the plane only about 0.5%, at most 1%. The numbers refer to the whole moisture range between fiber saturated and ovendry. In service the moisture contents vary less and involve correspondingly lower movements.

In thickness, fiberboard lacks stability, swelling and shrinking like particleboard, even more than solid wood perpendicular to the grain. In swelling-shrinking cycles, folded and wrinkled fibers straighten out and compression *springs back*. Some fiberboard not only swells in thickness more than solid wood, it also tends to swell faster because capillary spaces between fibers serve as relatively wide pathways for penetrating and evaporating water and water vapor. Porous light board swells and shrinks especially fast.

To retard moisture absorption and swelling, manufacturers add water-repellents (*sizers*), which repel liquid water much more than they do water vapor. Added synthetic resin binders improve the dimensional stability too; applied in small amounts they enforce bonds between fibers at some points, creating a sort of network which restrains swelling but still allows moisture to enter. The binders therefore retard swelling and shrinking more than water absorption.

Board Strength and Special Additives

The horizontal orientation of the fibers strengthens the panel in the plane at the expense of strength perpendicular to the plane, improving particularly its flexural properties. On the other hand, because of this orientation, surface layers tend to peel off and edge screws may split the center. The fibers do not align exactly at random in all plane directions, a majority line up in panel length as the slurry flows, with the effect of slightly greater lengthwise strength compared with strength widthwise.

Fiberboard strength and stiffness decrease with increasing moisture content up to fiber saturation, as is the case with solid wood and other wood-base panels. Exposure to humidity and water weakens the panels at least temporarily. This and swelling are the reasons why water absorption serves as an important board quality criterion. Adding sizers and binders naturally retards the loss in strength and stiffness.

Some fiberboard contains fire-retardants and preservatives for resistance to fire, decay, and wood-destroying insects. Most manufacturers dissolve the additives in the slurry to be precipitated on the fibers with alum, in this way achieving uniform distribution within the board. Surface coatings actually cost more and are less effective. Fiberboard which contains phenolic resin as binder is fairly durably without added preservatives and fire-retardants.

6.5. INSULATING BOARD

The standard *ASTM D 1554* (*Terms Relating to Wood-Base Fiber and Particle Panel Materials*) calls this product *structural insulating board.* According to *PS 57-73* (*Cellulosic Fiber Insulating Board*) it has densities from 10 to 31 lb/ft^3. Fiberboard of higher density falls into the category of *hardboard,* while below 10 lb/ft^3 it is *semirigid fiber insulation board.* The semirigid type resembles mineral insulation blankets and finds use as heat insulation in refrigerators, truck bodies, and so forth, where it retains its shape better than blankets.

In the insulating board manufacturing process, screens and rollers below and above the mat of pulp squeeze water out, compress the damp mat, and equalize thickness everywhere before the panel dries in air tunnels, without hot-pressing between platens. The screens and rollers cannot densify far beyond 31 lb/ft^3 (0.5 g/cm^3); hence this borderline between *insulating board* and *hardboard.*

Thermal Insulation and Strength

In buildings insulating board reduces loss of heat in winter and retards penetration of heat in summer. For reasons explained in Section 8.1, the lightest boards insulate best. The influence of density itself is small however, much smaller than in solid wood. Builders choose a board as light as strength allows because light board contains less mass and costs less for the same thickness. In other words, a given amount of fiber insulates better in the form of a thick light board than a thin dense board.

Interestingly, fiberboard insulates better than solid wood of the same density, a fact that is explained by discontinuities in fiberboard: heat flows across the board (as it does across solid wood) mainly in fiber walls around the insulating air spaces, but in solid wood these detours are much shorter than in fiberboard. The difference between wood and fiberboard diminishes in their lower densities where heat conductivity in the air spaces—the same for both materials—becomes more important. The ratio of heat passing through insulating board as compared to that passing through solid wood is 0.5 when both have a density of 31 lb/ft^3, and 0.8 at a density of 10 lb/ft^3. The R-value of ½ in. insulating board averages 1.25 (Section 8.1).

Insulating board insulates less than batts and blankets of glass fiber and mineral fiber (Table 8.1), but it resists mechanical force and provides firm even surfaces, as needed for sheathing of wood-frame houses for example. The sheathing confines the wall and prevents frame racking. Some panels serve as a nailing base for shingles. Dense board naturally tends to be relatively strong and stiff. Strength is the reason for using a board denser than 10 lb/ft.^{3}.

Uses, Types, and Dimensions

Insulating board is produced in two general qualities, *interior* and *sheathing*. Interior-quality boards are for where the panel remains dry. The sheathing quality contains a water-repellent that may also serve as binder; typically the manufacturers add asphalt to the fibers and/or coat the board with asphalt. Boards of this kind give sheathed buildings a familiar black appearance; *Celotex* is a well-known brand. Boards of both qualities, interior and sheathing, are produced for specific purposes, mostly for construction where 80% of all insulating board is used.

For boards that remain visible builders prefer a light color. *Factory-finished board* has a white paint coat, which usually includes a fire-retardant, and is intended for interior walls. It insulates better than gypsum wallboard but is inferior in hardness.

More than 20% of insulating board winds up as 1.5 to 3 in. thick *insulating roof deck* consisting of several bonded layers. The deck is intended for use without additional insulation; it insulates adequately in some parts of the country.

Acoustical insulating board, usually supplied as factory-finished tiles, has a fissured, felted-fiber, slotted or perforated surface pattern to minimize sound reflections and absorb sound. As ceiling and wall cover it minimizes noise originating in the same room. By contrast, brown *sound-deadening board* in wall assemblies is supposed to reduce sound transmission from room to room, a task for which no particular surface is needed (see Section 7.3). Acoustical and sound-deadening types conserve heat in addition to their acoustical effects.

Insulating board is marketed in a wide variety of lengths, widths, and thicknesses. For covering large areas, builders prefer the large but still convenient 4 × 8 ft panel. Quite common are also 1 × 1 ft tiles which give ceilings a varied appearance; many are of the acoustical type, but with a particular surface pattern chosen for appearance. Thickness varies: wallboard and some *shingle backer* types are only ³⁄₈ in. whereas all other types measure at least ½ in., as needed for thermal insulation.

6.6. HARDBOARD

Hardboard is a *hard, relatively thin fiberboard, manufactured by consolidating interfelted mats of fibers under heat and pressure in a hot-press to a density of 31 lb/ft³ or greater.* Heat and pressure reactivate lignin as a binder, which makes the board stronger than its density would lead one to expect in comparison with fiber insulating board.

Some people call hardboard *Masonite*; others apply the term *Masonite* to ⅛ in to ¼ in. dark hardboard. *Masonite* is a brand name which, strictly speaking means *board produced by the Masonite Corporation,* the oldest and best known hardboard manufacturer in North America. The corporation's first and for some time only panel was the dark hardboard; but Masonite has by now a widely diversified line of fiberboard products.

Color, Strength, and Density

The press platen heat of more than 380°F is what deepens the color of the board to greyish brown and dark brown. The fibers have darkened already to some extent during defibering. Since in contrast to interior-quality insulating board a dark color does not matter in hardboard, manufacturers take full advantage of the softening effect of heat and moisture, heating the chips liberally also to activate lignin as a binder. The *Masonite* defibering process causes darker colors than defibering in attrition mills: chips are steamed at 550°F in a strong vessel (*gun*) under the high pressure of 1000 psi, which when a quick-valve is suddenly opened explodes the chips to fibers.

The *ASTM Standard D 1554* defines *medium-density* hardboard from 31 to 50 lb/ft³ and *high-density* hardboard above 50 lb/ft³. *PS 58-73* (*Basic Hardboard*) disregards densities and classifies by strength and water resistance, the properties more significant to users, leaving it up to manufacturers which density will match the performance required. In reality, the requirements of the *PS 58-73 Industrialite* class (Table 6.4) are met by some but not all medium-density hardboard. The *Service* class has densities around 55 lb/ft³ and falls already into the high-density group, while *Standard* class densities average 60 to 65 lb/ft³.

Hardboard should be densest in the two surface layers, which contribute much more to flexural properties and hardness than does the core. The manufacturers achieve dense surfaces by closing their hot-press rapidly, so as to reach the desired panel thickness and maximum pressure when the surfaces are already hot and plastic but the core is still cold and stiff. The hot surfaces yielding to pressure become much denser, and cause pressure to drop before the heat wave can reach the panel core, which therefore consolidates less.

Table 6.4. Strength of Hardboard[a]

Class	Modulus of Rupture (psi)	Tensile Strength (psi)	
		Parallel to Surface	Perpendicular to Surface
Tempered	7000	3500	150
Standard	5000	2500	100
Service-tempered	4500	2000	100
Service	3000	1500	75
Industrialite	2000	1000	35

[a]Minimum averages per panel according to PS 58-73.

The manufacturers *temper* many hardboard panels in a second heat treatment to improve board properties. In North America the industry roll-coats both surfaces with oil and *bakes* the oil, both for deep penetration and to *dry* (oxidize) it. Europeans temper without oil, simply pressing the board once more but at a temperature higher than that of the first pressing. Tempering improves moisture resistance and increases bending strength as well as tensile strength, but it adds brittleness.

Screen-backed and Smooth Hardboard

The wire mesh screen on which the wet mat is formed remains under the panel in the hot-press; it permits steam to escape along the wires but leaves a distinct *screen-backed* impression (*weave*) on this *smooth-one-side* (S1S) board's *reverse* surface. For some uses the manufacturers press *smooth-two-side* (*S2S*) board, either from a dry mat between smooth platens or from a wet mat between cauls perforated with slots or holes, which leave small ridges or circular protrusions at the board surface to be sanded off. Dry mats are obtained either by drying *wet-felted* mats before pressing, or by *air-felting,* that is, drying fibers in air and letting them fall like fluffy brown snow.

Flexible wet fibers interfelt better and develop stronger natural bonds than do stiff dry fibers. Other conditions being alike, wet-pressed board (all *S1S* and some *S2S* board) has higher strength and water resistance than dry-pressed *S2S* board. The weak bond of the dry type can be compensated for, however, by adding synthetic binder and tempering. Where binder has to be added for exceptional strength and water resistance anyway, pressing dry air-felted mats seems quite practical, because spraying binder on dry fiber is more efficient than adding binder to a thin slurry whose water drains some away. Good air-felted medium-density board contains 7 to 9% binder.

Special Kinds

The industry processes hardboard further for specific uses. Perforated board (*pegboard*) is best known for this category, featuring closely spaced holes for various purposes, especially for hooks supporting all sorts of items. Pegboard can be made of any hardboard. The stronger classes give better service holding heavy items, but where the holes serve solely for ventilation or decoration (as in TV back and space divider panels) the weak classes are strong enough. Likewise, perforated board under ceilings with sound-absorbent material sandwiched between common hardboard and the ceiling requires little strength. Holes for such acoustical purposes must comprise at least 15% of the board area, which amounts to much more than the 1/8 in. holes per 1 in. of common pegboard.

For hardboard paneling, the cheapest kind of wall paneling, manufacturers press or score grooves into the surface to imitate narrow boards and to provide lines for nails; they imprint the grain of popular cabinet woods by means of engraved cylinders (Section 14.9). *Textured* board features decorative surfaces with patterns of ceramic tiles for bathrooms, of rough sawn lumber, bricks, leather, basket weave, or etched wood—all made by means of patterned press cauls.

Paneling leaves the factory completely finished, whereas siding may be only primed, that is, coated with primer to provide a surface receptive to paint. Other panels are marketed in painted, enameled, or overlaid surfaces.

Medium-Density Hardboard, Medium-Density Fiberboard, and Laminated Paperboard

Medium-density hardboard as defined in *PS 60–73* (*Hardboard Siding*) has been developed specifically for house siding; it is offered as lap and as panel siding. Lap siding yields a pattern of horizontal overlapped thin planks, which may be either embossed or smooth on the face. Builders fasten the 1 ft wide planks with concealed *hook-in* attachments provided by the manufacturer, or else they use nails. Panel siding, commonly 4 × 8 ft in size, forms a flush surface; besides smooth and embossed versions, the industry makes grooved panels to simulate board and batten. The siding runs 1/4 in to 7/16 in. thick and needs relatively little strength; for the thick kind the standard requires a modulus-of-rupture minimum of only 1800 psi.

Medium-density fiberboard (MDF), a sort of particleboard consisting of fibers, or a fiberboard-particleboard hybrid, is an interior product, urea-bonded, which lacks the moisture resistance of hardboard. The furniture industry uses MDF as a fine-structured core which can be machined to smooth edges and which holds screws fairly well.

Laminated paperboard is not really a fiberboard either, but resembles medium-density hardboard. Paper mills make it by combining plies of paperboard (typically a paper 1/16 in. thick and known as *cardboard*). Laminated paperboard comes 1/8 to 3/8 in. thick and finds uses similar to hardboard, as interior wall and ceiling finishing in mobile homes, as sheathing, and as lap siding with protective overlays.

6.7. COMPARISON OF PLYWOOD, PARTICLEBOARD, FIBERBOARD, AND LUMBER

Until several decades ago, lumber was used to cover large surfaces of roofs, walls, and floors which in modern structures are sheathed with plywood, particleboard, and fiberboard. Nevertheless lumber is still an alternative to panels. All four products are interchangeable in many other products also, including furniture and wall paneling. Which one of the four is most practical for a particular purpose depends on properties and costs. In structural uses, mechanical properties and prices are crucial, and shall be compared first. Dimensional stability is not discussed here since it has appeared earlier.

Strength and Costs

A great variety within each of the four products renders comparison difficult. Particleboard and fiberboard are pressed to desired densities and strengths, while lumber and plywood properties are less independent of natural wood and include defects. To make the four products comparable, Table 6.5 lists mechanical values for a density of 37 lb/ft^3, at which *fiberboard* is *hardboard.* Particleboard and hardboard values are actually chosen for the density **range** 37 to 50 lb/ft^3 in favor of these panels. Variability and defects of solid wood are taken into consideration as indicated in the footnote of the Table.

In modulus of rupture (MOR) and modulus of elasticity (MOE), solid wood is naturally superior in fiber direction to the lengthwise direction of crossbanded panels. On the other hand the panels, in direction of width—being roughly as strong that way as lengthwise—are much stronger than solid wood perpendicular to its grain. The 260 psi listed for pine amount to little more than 1/10 of the lowest panel MOR, 2400 psi. This explains why lumber edges break off much more easily than panel edges.

Considering the average of all directions, plywood as a reconstituted panel retains the strength of solid wood but particleboard and hardboard fall behind, especially perpendicular to the plane. When man disintegrates wood to small pieces and reassembles pieces into panels, he does not restore wood's original strength. In hardness, however the four products differ little. Plywood in this regard is assumed to equal solid wood, though crossbanding may harden it somewhat.

In price per unit volume, low grade lumber must be cheapest since its source—low grade timber—serves also as raw material for particleboard and fiberboard, whose manufacture costs more than sawing lumber. Whether particle-

Table 6.5. Comparison of Lumber and Wood-Base Panels[a] in Strength and Elasticity

Material	Modulus of Rupture (psi)	Modulus of Elasticity (psi)	Tension Perpendicular[b] (psi)	Side Hardness (lb)
Shortleaf pine[c]	8250	1020,000	260	400
Sweetgum[c]	7900	960,000	430	500
Shortleaf pine plywood	4250[d]	540,000[d]	200[e]	400
Particleboard type 1B2	2400	400,000	60	500
Hardboard	3000[f,g]	440,000[h]	35[i]	450[f]

[a] At 9 to 12% moisture content and 37 lb/ft^3 density.
[b] Perpendicular to grain and to the plane, respectively.
[c] Averages of clear straight-grained specimens, reduced for variability so that 5/6 of all individual values are higher, and reduced for defects by additional 25%.
[d] Average of solid wood in fiber direction and perpendicular.
[e] Assumed 3/4 of the value of solid wood.
[f] Minimum required for siding according to PS 60-73, and usually achieved in the medium-density type (31 to 50 lb/ft^3).
[g] For the relatively strong 1/4 in. thick panel.
[h] Unpublished measurement by W. C. Lewis.
[i] Minimum required for the Industrialite Class according to PS 58-73.

board or hardboard is lower priced depends on thickness: panels ½ in. and thicker cost less in particleboard, whereas panels ¼ in. and thinner cost less in hardboard. In the grade *A-D* and better, *Construction* plywood prices exceed those of particleboard and hardboard. Particleboard contains more binder but the particles cost only one quarter as much as veneer, and production of particleboard is more mechanized than that of plywood.

In price per surface area, hardboard as relatively thin panels comes cheapest. Veneer actually costs less but does not have sufficient transverse strength in the width direction. Where very little transverse strength is needed, for bushel baskets for example, veneer remains competitive.

Factors Other Than Strength and Costs

Since on volume basis some lumber costs less than panels, the question arises why builders do not use lumber for subflooring. They do not in the first place because narrow, dimensionally unstable lumber boards require more time to install than wide panels. Also, even tongued and grooved lumber joints are weak and hardly transfer point loads from board to board; hence lumber has to be thicker than plywood, which in turn lacks smoothness in the cheap grades. Builders equalize cheap plywood subflooring and bridge panel joints with particleboard or fiberboard underlayment, at the same time reinforcing the floor.

In comparison with cheap particleboard and hardboard, plywood remains competitive only where appearance and strength count. But even there particleboard may be the better choice, since in bending-stressed panels increased thickness efficiently compensates for inherent weakness: particleboard with half the strength, for example, needs less than twice the thickness of plywood to carry the same load.

Adaptation to Particular Needs

All manufacturers of lumber and wood-base panels have options with regard to size and properties of their products. Particleboard and fiberboard however can be produced in a much wider variety than plywood. In turn, plywood's options exceed those of lumber. Sawmills can choose how to cut the log, particularly how thick the lumber shall be, but the log diameter limits board width and the wood has to be accepted as is, with all particularities and defects. Plywood manufacturers engineer the panel through selection of the veneers, arrangements of the veneers in the panel and the glue type. Particleboard options are still more numerous particularly through size, shape, and orientation of the particles.

So far the number of options seems to be greater where the constituent pieces are smallest, or in other words, the number of options multiplies with the degree of wood disintegration during manufacture. But this rule does not extend to fiberboard, since the pulp hardly varies, leaving only a choice of board density, density gradient and additives.

The options enable producers to adapt the panel to consumers' needs, and to manufacture panels with particular features for special uses. This has contributed to the rapid growth of particleboard, fiberboard and plywood. The various lumber grades serve a similar function but are not so precisely suited to specific uses; instead the consumer accepts defects to save costs.

7

Acoustical Properties of Wood

This chapter concerns violins and construction. A *building material* book naturally deals mainly with construction. I shall discuss the violin first, because it is the best known of those musical instruments made almost exclusively of wood. In the other stringed instruments wood serves the same purpose.

7.1. THE ROLE OF WOOD IN VIOLINS

When plucking a stretched string, you hear a faint note caused by the vibrating string. The note is not loud because the string's small surface transmits little vibrational energy to the air. From the relatively few air molecules set in motion, the vibration disperses to so many others that the sound-energy is dissipated and air vibrates in the ear only slightly.

In the violin, the bridge under the string transmits string vibrations to a thin belly plate, which vibrates with the string and transmits these vibrations in turn via the sound post to a thin back plate. As belly and back surfaces are large, their vibrations engage many air molecules, rendering the note distinct and loud: the violin body amplifies.

Since violins are made from wood (and any other material is unthinkable among musicians), the question arises, what justifies this preference. The violin body should move immediately with the string and it should do so intensely. This requires material easily set in motion, that is, of low mass but adequate strength. Furthermore, belly and back plates should swing back after being pushed out of normal position by sound waves; they should be elastic and stiff, having a high *modulus of elasticity* (MOE). The material needs a large *MOE-to-density* ratio.

Wood really stands out in this respect; its *MOE-to-density* ratio is about the highest of all materials. Among the various wood species the ratio of spruce ranks highest. Spruce indeed is the preferred wood species for stringed instruments. Whether a high MOE and low density are desirable for acoustics of buildings is another question.

7.2. SOUND ABSORPTION IN BUILDINGS

Acoustical problems in construction are of two basically very different kinds. One is *room acoustics,* that is, *control of sound that originates in the same room,* making it distinctly audible, amply loud, and with as faithful a replication of the source as possible in all locations within the room. The other is *sound isolation,* the *control of noise transmission from room to room.* This section deals with room acoustics, the next section with sound isolation.

Significance and Mechanism of Sound Absorption

Room acoustics depend on absorption of sound by walls, ceiling, floor, and furnishings. Objects either absorb or reflect sound waves. In a room with insufficient absorption, walls and other surfaces reflect sound repeatedly, the waves bouncing back and forth in every direction, literally filling the room. This multiple reflection causes a build-up, sound becoming unnecessarily loud in distant parts of the room. When the source stops, waves continue to travel back and forth diminishing only gradually, so that listeners hear the sound prolonged and drawn out as reflected waves strike the ear in rapid succession. In other words, there is too much *reverberation time,* the *time required for a sound to diminish to a certain fraction.* Adequate absorption makes each sound distinct and the room quiet, affecting sound *persistance* naturally more than sound *intensity* while the source is still producing sound.

Sound absorbing materials are porous, permitting air molecules to vibrate in and out of the surface pores. Within the absorbent material, air molecules vibrate in narrow passageways and cause friction, with the effect that sound energy dissipates into heat; ultimately all sound becomes heat. For absorption the pores must be barely large enough for blowing air through. Smaller pores reflect sound, whereas larger pores permit air to move back and forth without friction.

Absorption by Wood and Wood-Base Materials

The air that enters lateral wood surfaces passes through cell wall pits which are too small for significant sound absorption. Wood like most substances belongs to the category of *sound-reflecting* materials; it reflects 90 to 95% and absorbs only 5 to 10% of the sound. Rough wood absorbs more than planed wood but the difference is trivial.

Acoustical materials (*absorbents*) absorb at least 50%. One of them is *acoustical tile,* basically fiber insulating board with holes or fissures that comprise at least 15% of the surface. Common fiber insulating board absorbs hardly more than solid wood. Extremely rough, porous particleboard may qualify as an absorbent, while common particleboard and hardboard do not. Wood-fiber felt absorbs no more than acoustical tiles of the the same thickness; however, most commercial felts are thicker and correspondingly more effective than the tiles.

Sound absorption depends heavily on sound frequency. Wood-base panels absorb high sounds much more than deep ones and have absorption maxima around 1000 to 2000 Hertz.

Wood end-grain surfaces are rarely exposed and their absorption has not been explored, at least not to this author's knowledge. It can be assumed that end grain absorption exceeds that of lateral surfaces, because the exposed cell cavities are many times larger than the pits. However, after a sound wave has penetrated the surface, it makes no difference where it came from. The cell-wall substance itself plays an insignificant role in the absorption process, only the geometry of the pores counts. Wood has no inherent absorbing capability.

Absorption of Paneling

The forementioned considerations apply to wood paneling that is firmly attached to a rigid backup surface, without air space between panel and wall. Panels mounted an inch or so from the wall can vibrate, forming an absorbing unit along with the trapped air, the latter constituting the primary sound-absorbing component of the unit. In this case wood takes part in absorption just as any flexible, impermeable material of the same mass per unit area would do.

Absorption of paneling with an air space underneath varies with the resonance frequency of the assembly. Paneled rooms have reverberation times longer in some ranges of frequencies than in others. These rooms tend to absorb mainly low frequencies (deep sounds), destroying the warmth of music and lending a shrill, unpleasant *acoustic.* Fortunately, absorption in the various frequencies can be controlled through the depth of the air space and by filling the space with absorbant material.

7.3. SOUND TRANSMISSION

How loud one hears noise in an adjoining room depends on sound transmission through the partition. Sound waves striking the partition set it into vibration, presumably affecting its opposite surface also, which in turn causes vibration of adjacent air and radiates the vibrations away as transmitted sound. In brief, partitions transmit sound by means of surface vibration to the receiving room.

Transmission as Function of Mass

Heavy walls transmit little. A wall of infinitely great mass would vibrate not at all, whereas an extremely light partition vibrates even from a whisper. Low wood density combined with high stiffness favors vibration, as shown in the violin, but in contrast to violins, these vibrations are undesirable in structures.

Wood-frame partitions surfaced with gypsum wallboard do relatively well at isolating sound, though gypsum is only a little denser than most wood species. Wallboard however is thicker than plywood or hardboard paneling and has therefore more mass. Partitions consisting of wallboard over fiber insulating board (*sound-deadening board*), applied to both sides of the studs, insulate much better than wallboard-stud-wallboard partitions. The added insulating board works because it differs from wallboard, and through increased mass besides.

Solid-core doors transmit less sound than hollow doors, again due to their higher mass. But some sound passes by, through air leaks along the edges; hence adding gasketing may reduce sound transmission more than replacing hollow-core doors with solid ones.

Paneled Partitions

The panels of partitions vibrate inbetween studs, sound waves press them in and out (Fig. 7.1). If the sound frequency comes close to the panel resonance frequency (flexural wave length), the panel vibrates intensely and transmits most of the sound to the adjoining room, while transmitting little sound in other frequencies.

Partitions with wallboard on both sides of studs transmit vibrations either as airborne sound through the wall cavity, or as *structure-borne* vibration through the studs themselves. The latter process works as follows: an incident sound wave incites a flexural vibration in the source-side panel, forcing studs into torsional vibration; in turn the studs excite a flexural wave in the panel of the receiving side. Studs effectively transmit vibrations whenever they are firmly connected to both skins. Spongy connections reduce transmission of this kind, so to this end builders insert a resilient material such as fiber insulating board between stud and panel, connecting everything with adhesives instead of nails.

Thin, no-load-bearing metal studs transmit little sound by torsional vibration, because the two resilient legs of their U-shape vibrate independently of each other. An even greater improvement would be the double leaf design:

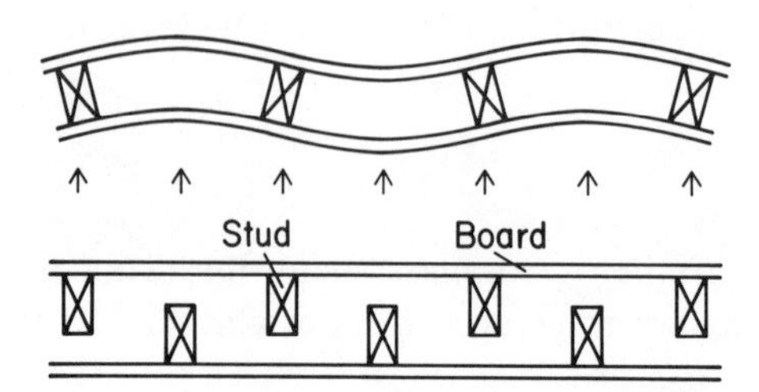

Figure 7.1. (Top) Structure-borne sound transmission through stud torsion; the arrows indicate an incident sound wave and torsion, respectively. (Bottom) Wall with staggered studs. (Adapted from Schultz, 1969.)

two rows of staggered studs with the panels on one side attached to one row and the panels on the other side to the other (Fig. 7.1).

Impact Isolation

Footfalls or other impacts on the floor above a ceiling are quite loud in many houses. The floor-ceiling structure being poorly insulated, transmits the impact readily to the room below. Generally, impact-sound isolation requires a heavy, non-resonating design, a feature not possessed by wood structures. Wood-joist floors are easy to set into motion and represent poor sound insulators, but carpets and pads below the carpet serve as a sort of cushion which absorbs sound before it can reach the floor.

For impact isolation under hardwood-finish floors, builders sandwich a layer of fiber insulating board between subfloor and underlayment. In this *floating design* the insulating board prevents impact transmission to the ceiling below.

8

Thermal Properties and Fire Performance of Structural Wood

This chapter explains how wood responds to daily and seasonal changes in temperature as well as to daily temperature differences, how it reacts when close to hot items, and how it holds up in burning buildings. The relationship between strength and temperature has already been covered in Chapter 4, but only for moderate temperatures whose effects are temporary. I shall include wood's fire performance as regards construction, while combustion in the context of heating with wood stoves and fireplaces follows in Chapter 9.

8.1. PHYSICAL THERMAL PROPERTIES

Properties of this kind deserve attention for a variety of reasons. To build sensibly we need to know wood's *thermal expansion,* that is, how wood dimensions change with temperature. *Thermal conductivity,* in construction expressed by the *R value,* determines the winter heat loss of buildings and the summer heat which enters through walls and window frames. *Specific heat* along with thermal conductivity affects how warm wood feels on touch and how fast the temperature rises inside heat-exposed material, such as wood in fire.

Thermal Expansion

Like the rising and falling mercury in the capillary tube of thermometers, most materials expand on heating and contract on cooling. Wood reacts basically the same way, though shrinking and swelling occurring simultaneously may overshadow thermal dimensional changes. Thermal contraction should not be confused with contraction due to thermal degradation (Section 8.2), which is permanent and not reversible, does not result from temperature changes but occurs simply at high temperatures. We express thermal expansion by the *coefficient of linear thermal expansion,* in units of length per unit of length and degree change in temperature (in./in. °F).

Just as with expansion due to moisture, wood expands with heat least of all in its fiber direction, where at 12% moisture content the coefficient of expansion has the low order of magnitude of 6×10^{-6} in./in. °F. Steel, window glass, concrete, and bricks expand by similar amounts, aluminum and plastics by much larger amounts (Table 8.1). To compare thermal expansion with swelling and shrinking, consider an exterior wall stud and assume its temperature to fluctuate 100°F, from an extreme of −10°F in winter to 90°F in summer. This produces an overall thermal expansion of 6×10^{-4} in./in. or 0.06%, less than 1/16 in. in the 8 ft stud. We may assume that the rising temperature causes the humidity to drop and the wood to dry from 15 to 5% moisture content. The resulting shrinking is about 0.1% (Section 3.6)—hardly more than the thermal expansion: stud length thus remains fairly constant through different seasons.

Thermal expansion perpendicular to the grain is higher tangentially than radially for most wood species, moisture contents, and temperature ranges, the ratio of tangential to radial expansion being about 1.3 to 1 on the average. In the hygroscopic range, the expansion coefficient increases with an increase in moisture content, up to a maximum near 20% moisture content, dropping above fiber saturation to a level lower than in the ovendry state. Toward high temperatures the coefficient increases slightly. Denser wood species generally expand more than light species, but the variations among species of equal density is so large that the influence of density is negligible in the common specific gravity range from 0.3 to 0.7.

Our native wood species expand transversely 25×10^{-6} in./in. °F, on the average, at room temperature and 12% moisture content. Comparing this with shrinkage, again consider the example of the stud, which during a temperature rise from −10 to 90°F expands about 0.2 or 0.3%. The average of tangential and radial shrinkage when drying from 15 to 5% is near 2%, or eight times higher than the thermal expansion: transverse shrinking and swelling overshadow thermal expansion.

Specific Heat

Specific heat is the amount of heat required to raise the temperature of a unit mass of the substance by one degree. Our units of heat are by definition related to the specific heat of water: one British thermal unit (Btu) raises the temperature of one lb of water one degree Fahrenheit, while one kilocalorie (kcal) raises the temperature of one kilogram one degree centigrade. Hence,

Table 8.1. Thermal Properties of Materials at Room Temperature (Wood and Wood-Base Panels at 12% Moisture Content)

Material	Density (g/cm^3)	Thermal Expansion (in./in. °F)	Thermal Conductivity (Btu in./hr ft^2 °F)	Specific Heat (Btu/lb °F)	Thermal Diffusivity ($in.^2$/hr)
Douglas fir, parallel grain	0.54	6×10^{-6}	2.40	0.39	2.19
Douglas fir, across grain	0.54	21×10^{-6}	1.00	0.39	0.91
Northern red oak, across	0.71	32×10^{-6}	1.25	0.39	0.87
Redwood, across	0.45	23×10^{-6}	0.85	0.39	0.93
Fiber insulating board	0.29	6×10^{-6}	0.40	0.39	0.68
Charcoal, pine, across	0.26	---	0.69	0.23	2.20
Dead air	0.0012	---	0.18	0.24	118.4
Polystyrene, expandable	0.032	35×10^{-6}	0.24	0.32	4.5
Mineral wood, glass type	0.04	---	0.27	0.20	6.5
Glass, window type	2.50	5×10^{-6}	6.00	0.18	2.6
Gypsum board	0.80	---	1.13	0.26	1.04
Brick	1.93	5×10^{-6}	5.00	0.22	2.26
Concrete	2.27	8×10^{-6}	12.60	0.16	6.84
Polyvinylchloride, rigid	1.40	38×10^{-6}	1.13	0.24	0.65
Steel	7.85	8×10^{-6}	324	0.10	79.3
Aluminum	2.70	14×10^{-6}	1080	0.23	334.3
Water	1.00	---	4.03	1.00	0.78

the specific heat of water is unity in both systems, British and metric, and 1 btu/lb °F equals 1 kcal/kg °C.

All wood species exhibit the same specific heat under the same conditions, that is, 0.29 Btu/lb °F at room temperature and in the ovendry state. Since water's specific heat is more than three times higher, the specific heat of wood goes up as its moisture content increases, reaching 0.5 Btu/lb °F at fiber saturation. The specific heat also increases with rising temperature, though not as much. Dry wood's specific heat approaches 0.4 Btu/lb °F at the boiling point of water, while that of moist wood increases even less, since water's specific heat is nearly the same at all temperatures above freezing.

Thermal Conductivity

Heat transfer occurs in three ways: radiation, convection, and conduction. Thermal conductivity, concerning only the latter, is a measure of the rate at which heat is conducted through a material.

Measurement and Units To determine thermal conductivity, take a panel of thickness t, heat one surface to a high temperature h, simultaneously cool the opposite surface to the low temperature ℓ and measure heat input and/or output. The temperature gradient $(h - \ell)/t$ is what drives heat through the panel. Thermal conductivity is expressed in British thermal units (Btu) per hour (hr) per square foot of panel surface (ft^2) per temperature gradient (°F/ft or °F/in.), and is written in the combined unit

$$\frac{\text{Btu}}{\text{hr ft}^2\ {}^\circ\text{F/ft}} \quad \text{or} \quad \frac{\text{Btu ft}}{\text{hr ft}^2\ {}^\circ\text{F}} \quad \text{or, abbreviated,} \quad \frac{\text{Btu}}{\text{hr ft}\ {}^\circ\text{F}}$$

Engineers prefer Btu/hr ft °F, or kcal/hr m °C, the corresponding unit in the metric system.

Thermal conductivity (k) is an inverse measure of the insulating value or resistance to heat flow. Commerical construction literature gives this resistance as R-value, which increases in proportion to the material's thickness (t):

$$R = \frac{t}{k}$$

[t (in.), k (Btu/hr ft^2 °F/in.)]. The common insulation blankets of fiberglass achieve R-11 if 3.5 in. thick, and R-19 if 6 in. thick. R is given without unit.

Influence of Density, Moisture Content, and Orientation The wood cell-wall substance's rate of heat conduction may be relatively low, but it is still much higher than that of dead (motionless, trapped) air. The *wall-to-air conductivity* ratio perpendicular to the grain is 12:1. Heat therefore flows in dry wood mainly along the cell walls, and conductivity increases with the wood's density. Since water conducts much more heat than either air (Table

8.1) or the dry cell wall, wood's thermal conductivity also increases with higher moisture content.

Due to the orientation of cellulose chains and the long straight path in cell walls, heat flows in fiber direction at least twice as fast as transversely. Longitudinal conductivity is perceptible in some items such as the handles of hot tools and floor joists lying at right angles to exterior walls.

Transverse conductivity plays however a much more important role; it determines home heat loss through window frames, exterior wall studs, wood siding, and ceiling joists. The rate is nearly the same tangentially and radially. The slightly higher radial conductivities are due to radially oriented ray cells and the arrangement of fibers in radial files (Figs. 1.1a and 1.9).

Significance Wood insulates very well (Table 8.1). Foamed polyurethane and styrofoam surpass it in this regard but are much lower in density and strength. Conductivities of foams equal that of dead air. Other air, such as in the space betweeen panes of thermal windows, insulates less well because it can move and transfer heat by convection as well as by conduction.

Vacuum is a better insulator than dead air, as demonstrated in thermos bottles. Vacuum however is unfeasible in building construction. In double-pane windows, for example, if the space between panes were to be evacuated, the glass would have to be extremely thick and divided in small sections to withstand atmospheric pressure. Thermos bottles retain heat also because the facing surfaces of their evacuated space are coated with a reflective substance to reduce radiant heat loss. This principle is occasionally applied to the air spaces of exterior walls, in the form of aluminum backed panels.

Wood's low thermal conductivity unfortunately contributes to its becoming charred by burning cigarettes. The heat remains isolated and raises the wood temperature to the level of thermal degradation. Aluminum backing foil under surface veneer prevents the damage by conducting heat away, distributing it over a large area, in this way keeping the temperature under the burning cigarette to a safe level.

Wood feels warm because it conducts little heat out of fingers, leaving skin nerves warm. What we feel is skin temperature, not the temperature of the wood.

Thermal Diffusivity

In a solid body suddenly exposed to heat, first the surface temperature rises. The surface conducts heat to the unseen layer below, where it is absorbed according to that layer's specific heat. The second layer in turn conducts heat to the third, and so forth. The rate of heat degree penetration is directly proportional to thermal conductivity (k) and inversely proportional to the heat capacity of each layer. The *heat capacity,* a sort of *specific heat based on volume,* is the product of specific heat (c) and density (D) expressed in units of Btu/ft^3, for example. Hence the degrees of heat penetrate in proportion to the ratio k/cD, which is known as *thermal diffusivity*; its dimension—in.^{2}/hr—is difficult to picture but makes some sense in the less abbreviated form °F/hr °F/in.^{2}, meaning change of temperature per hour per temperature gradient squared.

Significance and Magnitude Thermal diffusivity determines how long one can hold a tool whose temperature rises at the other end, for example when welding. In summer, the outside surface temperature of house siding reaches a high in the early afternoon, but it takes some time for the heat wave to reach the inside surface of the wall. Heat waves pass through wood-frame walls with insulation blankets between studs faster than through masonry walls of the same thickness, because the blankets are relatively high in thermal diffusivity. Since masonry walls are thicker, they heat up extremely slowly.

The thermal diffusivities of insulation blankets and plastic foam by far exceed those of wood. To explain this and to illustrate the difference between conductivity and diffusivity, consider two styrofoam coolers: one empty, the other filled with ice, both kept in a freezer. Taken out on a hot summer day, heat waves rapidly penetrate the styrofoam because of its high thermal diffusivity, raising the air temperature in the empty cooler. In the other the ice keeps for several days, because melting consumes much heat, all the styrofoam can conduct. Conduction of heat (Btu units) differs from the penetration of temperature (heat degrees, °F).

The low thermal diffusivity of wood is essential for doors and walls which isolate burning rooms. A properly constructed solid wood door will resist fire very well. It takes a long time for fire to pass through to the opposite surface, much longer than in hollow doors or doors with mineral wool cores. The thermal degradation discussed below contributes to this difference.

Thermal diffusivity is needed to estimate the temperature rise in heat-exposed material. One can calculate, for example, that the temperature in the center of a one-inch yellow birch board rises in a 220°F hot-press from 70 to 160°F in 10 minutes (min), and in 20 min to 195°F. In a 20 in. diameter green walnut log dropped into stirred hot water of 160°F, the temperature 2 in. from the pith and much farther from the end surface rises in 60 hr from 50 to 150°F. The calculations involve differential equations which are not explained here.

Influence of Density and Moisture Content Both thermal conductivity (k) and heat capacity (cD) increase with wood density and moisture content. Whether and how much thermal diffusivity (k/cD) will change as density and moisture content increase, depends on what increases more, conductivity or heat capacity; actually the latter does, hence dense species and wet woods are relatively low in diffusivity. Heat waves penetrate into dense wet wood more slowly than into light dry wood. However, the differences among the species and between dry and wet woods are small. Even the diffusivity of dry balsa, the lightest of all woods, is only two times higher than that of wet hickory, one of the densest native species.

8.2. THERMAL DEGRADATION

Strictly speaking, wood does not burn; heat first breaks it down chemically in a thermal degradation or *pyrolysis* process. Wood constituents decompose into pyrolytic substances—mainly gases and charcoal. The gases diffuse and stream out of the porous material to burn above the wood surface while charcoal burns *in situ*. Heat of combustion in turn promotes gasification and sustains the process. This section concerns pyrolysis only; Sections 8.3 and 9.2 deal with ignition and combustion.

Pyrolysis proceeds in overlapping phases. Reactions characteristic of the first phase extend into the second phase, while second phase reactions begin in phase one. The phases overlap in each tiny piece of wood and in each thin layer of uniform temperature. In thick pieces, pyrolysis reaches different phases simultaneously in different layers. For example, degradation in the surface layer of a post may cease entirely before the first heat wave and the first phase of degradation engulf its core. For simplicity's sake I shall disregard overlaps of time and place to focus on the major reactions, disregarding for now that products of different phases interact with each other.

Drying Phase

In wood exposed to the heat of fires, moisture evaporates and diffuses to the surface; above 212°F it boils, builds up steam pressure, and streams out of pores with a hissing noise. Free wood moisture begins to boil at 212°F, but steam pressure raises this boiling point. Bound moisture boils even at atmospheric pressure at elevated temperatures which rise with the remaining moisture content: to 214°F at 20%, 223°F at 10%, 243°F at 5%, and to 284°F at 3% moisture content. In rapidly heated porous wood the last traces of bound moisture escape when the temperature reaches 300°F. The conversion of water and bound moisture into vapor consumes large amounts of heat (Table 8.2). Hence, wood moisture holds the temperature of heat-exposed wood down, retards penetration of heat, and delays thermal degradation.

Rapidly drying wood checks and crackles from shrinkage and from steam pressure that bursts the tissue open, producing clefts through which the steam escapes. Steam pressure causes checks especially in wet wood of impervious species. The checks run in fiber direction, mainly in the weak radial plane, perpendicular to the direction of high tangential shrinkage.

Strictly speaking drying is not a phase of pyrolysis, but precedes it. Recooled dry wood reabsorbs moisture and returns to the original state, whereas real pyrolysis is irreversible: broken-up wood constituents do not reunite even if the fragments happen to remain together in the wood.

Slow Pyrolysis

This first phase of real thermal degradation shows in the odor of the escaping pyrolytic gases, loss of wood weight, and lowered strength. The gaseous

Table 8.2. Amounts of Heat (Btu/lb) That are of Interest for Pyrolysis and Burning

To raise the temperature of ovendry wood from 70 to 300°F	83
To raise the temperature of fibersaturated wood from 70 to 300°F and to dry it, based on weight when ovendry	450
To evaporate water at 212°F	970
Consumed for wood's slow pyrolysis, based on weight when ovendry	180
Released in wood's rapid pyrolysis, based on weight when ovendry	500
Heat of combustion of ovendry wood[a]	8,100
Heat of combustion of charcoal[a]	13,500
Heat of combustion of anthracite[a]	13,800
Heat of combustion of petroleum oil[a]	18,000
Heat of combustion of gasoline[a]	18,400
Heat of combustion of natural gas[a]	22,000

[a]Low heat value, that is not including latent heat of water vapor which can be gained by condensing the vapor.

compounds contain many hydrogen and oxygen atoms, so that carbon accumulates percentagewise and causes the wood remaining to look darker; but not all thermal darkening results from higher proportions of carbon.

It is not possible to pinpoint a certain temperature above which wood breaks down. At moderate degrees of heat the degradation progresses so slowly that changes can be detected only after long periods of exposure. It is a question of agreement, what length of time to consider. Furthermore, moist wood reacts at lower temperatures than does dry wood. Moisture accelerates the chemical changes and causes hydrolysis, a chemical reaction involving water.

Temperatures that occur in trees in nature can be assumed to cause no degradation; wood is designed to withstand natural degrees of heat. Some people believe though that in violins and pianos the elastic properties of wood change even at room temperature and lead to loss of *sound* after a few hundred years.

In moist wood heated above 160°F, insignificant decomposition appears in a day or two. Kiln drying of lumber causes slow degradation from about 160°F up. In originally air-dry samples exposed to 215°F, permanent changes show within a few days. The decomposition proceeds fairly steadily, slowing down gradually. At 300°F, for example, the loss in weight during the first two weeks diminishes little from day to day, averaging about 2% per day.

The first pyrolytic gases consist mainly of water vapor, traces of carbon dioxide, formic acid, and acetic acid. The vapor originates from *water of constitution* in hemicellulose and other wood compounds; *free* and *bound* moisture have escaped earlier, in the drying phase. The acids are what gives a characteristic smell to mills which heat wood for veneer cutting and wood bending.

Above 300°F wood temperature, the released gases form a smoke that consists mainly of suspended droplets of condensed water vapor. Around 400°F appear small amounts of methanol (commonly called *wood alcohol*), carbon monoxide, and other gases which become combustible only in higher concentrations, when the proportion of water vapor diminishes.

Rapid Pyrolysis

Up to 480°F the rate of degradation steadily increases with increasing temperature; then, in the phase of *rapid pyrolysis,* the process accelerates faster and the rate rapidly reaches a maximum, the wood losing much weight in a very short time. Beyond this peak, pyrolysis slows down, even when the wood temperature continues to rise. The degradation of spruce shown in Fig. 8.1 is typical, except that with other species the line's minor *bumps* appear at different temperatures or not at all.

Most of the rapid-pyrolysis gases are combustible; they include carbon monoxide, methane, and formaldehyde, but little water vapor or carbon dioxide. By reacting with each other and condensing, some of the gases form *tar,* a mixture of compounds which appears as a thick, sticky, brown to black liquid with a pungent odor. Some wood-pyrolysis tars develop within the wood and remain there, others form an *aerosol* (tiny droplets suspended in the atmosphere) and contribute to smoke in fires. The tar fraction is relatively large when wood is heated quickly.

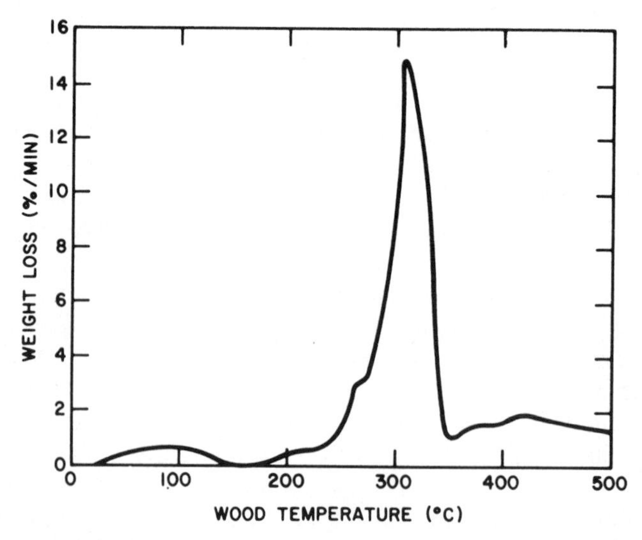

Figure 8.1. Thermal degradation of a small vacuum-dried spruce sample during heating under exclusion of oxygen at a rate of 18°F (10°C) per minute. (By N. Christoph.)

Chemical reactions are either *endothermic* (heat consuming) or *exothermic* (giving off heat). Under endothermic degradation the wood temperature cannot rise above that of the heat source outside it. Exothermic degradation causes self-heating; it raises the wood temperature above the surrounding temperature whenever the amount of generated heat exceeds the heat which dissipates to the environment. Exothermic pyrolysis promotes itself, while endothermic pyrolysis depends on an external source of heat to continue.

In fast-heating experiments, in which wood samples were heated above 900°F within a few hours under exclusion of oxygen, pyrolysis consumed heat until the wood temperature rose above 480°F. Then, in the rapid pyrolysis phase, the temperature continued to rise without further heat input and without any burning. Later, far beyond the peak of maximum rate of weight loss, the degradation reverted to the endothermic kind. Whether pyrolysis of wood maintained at moderately elevated temperatures for long periods of time is endothermic or exothermic is a question not yet clarified, though it has quite practical consequences for spontaneous ignition.

Charcoal Formation

Around 900°F pyrolysis comes to an end. The solid residue, *charcoal,* resists further degradation in the absence of oxygen; it consists mainly of carbon and weighs about one quarter of the original ovendry wood. This weight depends on the wood's content of carbon-rich lignin and on the heating rate—slow heating leaves one half, quick heating one eighth of the original ovendry weight as charcoal.

The loss of wood substance as volatiles involves contraction to one-half of original volume. The degrading wood contracts in all directions, not only transversely. This contraction determines the shape of the charred wood. Since deeper layers restrain contraction of the fast-degrading outer layers, the latter check in fiber direction and transversely (Fig. 8.2a). Finally the checkered char disintegrates into small cubical pieces.

Do not confuse *charcoal* with the *charcoal briquettes* marketed for outdoor cooking, which are made by pressing charcoal powder with cheap binders such as starch. Natural charcoal is so soft and brittle that it can be crushed almost with bare fingers, though it still retains the cell cavities and other structural characteristics of the wood species. Loss of substance leaves the cell walls extremely *open,* with a large internal surface for reaction with other substances, useful for example as an adsorbent in gas masks. Section 9.4 deals with combustion of charcoal.

Strength Loss and Charring Rate

The piece in Fig. 8.2b burned as part of a structure until the fire was suddenly extinguished. Note the sharp borderline between charcoal and wood. Outside this borderline, pyrolysis reached the rapid phase, in which generated heat

Figure 8.2. Charred lumber. (*a*) Lateral surface. (*b*) Cross section, most charcoal broken off.

accelerated degradation and drove it to the terminal stage of charcoal. Inside the borderline, the degradation progressed to different stages in different layers; hence the gradual change in color without sharp lines of demarcation. The wood is still essentially intact and able to carry a load; drying (caused by fire) partly compensates for the fire's degradation.

The *charring rate*, that is, the *rate of penetration of char,* serves as measure for strength loss. Wood chars in standard fire tests at a rate of roughly 1.5 in./hr; two-by fours degrade in about 40 min, but thick timbers retain strength for many hours. By contrast, steel and aluminum I-beams and U-shaped studs soften in fire very rapidly (Fig. 8.3); due to their extremely high thermal diffusivity and thin cross section, they heat up almost instantaneously and collapse long before timbers fail (Fig. 8.4). Likewise, wooden boxes protect

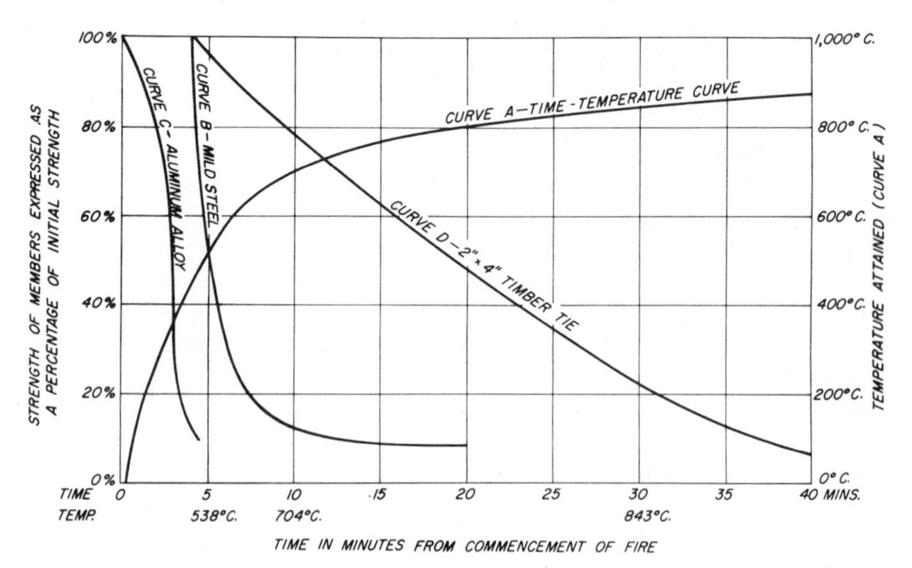

Figure 8.3. Strength loss of timber, steel, and aluminum in fire. (Photograph courtesy of U.S. Forest Products Laboratory.)

Figure 8.4. Lumber supports collapsed steel I-beams in a burned-out structure. (Photograph courtesy of U.S. Forest Products Laboratory.)

documents from fire much longer than metal boxes; wood may burn on the outside, but it takes more than half an hour for charring to reach the inside of 1 in. lumber or of wood-base panels.

The charring rate mentioned applies for the common construction lumber species Douglas fir and southern yellow pine. Lighter species degrade faster, denser woods slower, the reason being that denser woods have lower thermal diffusivity and especially larger masses of moisture. Also, larger wood-mass consumes more heat for endothermic pyrolysis, and the insulative dense charcoal burns off at a slower rate.

Lumber chars in fire at rates which change. Charring commences after the surface layer has heated up; it proceeds relatively quickly until charcoal begins to retard the penetration of heat. In a fire's first stage, the heat generated by exothermic degradation is consumed in evaporation of moisture and in endothermic degradation of deeper layers. But when the front of exothermic pyrolysis approaches the center of the piece, all the heat generated drives the temperature up and degradation accelerates. This is why the last third of timber *burns* at least as fast as the second third, even though combustion heat has a long way to go through the insulating charcoal layer to the core.

8.3. IGNITION

Wood pyrolysis products unite with oxygen under certain conditions. Oxidation as an exothermic reaction generates heat which converts more wood into combustible pyrolytic products, in this way feeding oxidation. The reaction progresses to rapid *combustion* or *burning* in the form of fire. *Ignition* is the *initiation of combustion.*

Ignition of Pyrolytic Gases

Pyrolytic gases ignite under three prerequisites. (1) The gases must be combustible. Water vapor and carbon dioxide, the first gases released from degrading wood, are oxides already and cannot burn, while formic acid and some other products of slow pyrolysis, as well as most rapid-pyrolysis gases and tars, can. (2) The combustibles and oxygen must meet in sufficient concentration. The first traces of carbon monoxide are too much diluted by water vapor and carbon dioxide for burning. Later, during rapid pryolysis, combustible gases may be too concentrated, forming an oxygen-depleted blanket at the wood surface and burning only above this blanket. (3) The mixture of combustible gases and oxygen must be hot, the degree of heat depending on the mixture.

The minimum temperature at which pyrolytic gases ignite *spontaneously* (without a pilot flame) has been reported at 806°F for formaldehyde, 1002° to 1382°F for methane, 1051°F for acetic acid, around 1085°F for hydrogen, and near 1200°F for carbon monoxide. Mixtures require at least the ignition temperature of formaldehyde (806°F), their lowest-igniting component.

For *induced* ignition with a pilot flame—a match, for example—the mixture's initial temperature makes no difference because it is the pilot flame which raises the temperature to ignition level. Induced ignition depends only on composition and concentration of the mixture, which in turn varies with the state of wood degradation. In the phase of rapid pyrolysis, above 480°F wood temperature, the gases tend to be sufficiently concentrated and to have the proper composition.

Spontaneous ignition and induced ignition differ very little essentially. The spontaneous kind may start with two molecules (one of them oxygen), which incidentally move much faster than the others, collide, and break into atoms that unite with atoms of the other molecule. The heat set free in this reaction raises the temperature of neighbor molecules, speeding their movement, triggering further collisions, and so on. The two molecules which react first are like a pilot flame, the only difference being that the molecular pilot flame is extremely weak and requires more favorable conditions than a real flame to be effective. This model illustrates that ignition depends on sufficient concentrations of both gases, since otherwise partner molecules stand little chance of colliding.

We know from burning firewood that some species ignite relatively easily, most notably the softwoods white pine, spruce, and fir; their easy ignition is explained by their high thermal diffusivity and the resulting rapid degradation of these light species, as well as by the presence of quite flammable compounds in resin. Pine tar ignites spontaneously already at 670°F. Flammable extraneous compounds may also be one reason why birch ignites so well, another being the type of bark (Section 9.5).

Low Temperature Ignition and Self-Heating

Through the decades numerous wood fires have been reported which started at temperatures as low as 250°F, even below 200°F. Joists that had been close to steam pipes or to heat ducts for months or years suddenly burned. Wood decomposition by thermophile bacteria had been observed to raise the temperature in piles of moist sawdust and of wood chips until heat killed off the microorganisms at around 180°F; then surprisingly the temperature continued to rise, the pile of wood particles heated itself and finally burned. In these reported cases the material was somehow shielded or insulated, so that little heat dissipated to the surroundings and the pyrolytic gases could accumulate. Frequently an influx of humid air accompanied self-heating. Fiberboard appeared to self-heat more than other wood-base panels and solid wood.

It is not known how wood ignites spontaneously at low temperatures. The ignition probably starts with charcoal, whose formation does not require high temperature. Wood exposed to moderate heat around the boiling point of water gradually chars, in many months of exposure. The resulting charcoal is highly reactive; researchers have observed its spontaneous ignition at 300 to 480°F. Under special conditions, when sufficient oxygen is available and heat generated by a few incidental oxidations of carbon atoms does not dissipate, the ignition point may be much lower. Charcoal plays a similar role when people rekindle a fire: blowing accelerates combustion of the charred wood surface, which in turn ignites pyrolytic gases.

8.4. IMPROVED FIRE CHARACTERISTICS

Wood withstands fire fairly well, but for some purposes its fire resistance is insufficient. The industry offers lumber and wood-base panels which are treated with *fire-retardant* chemicals for the following specific goals:

Retarding Pyrolyis for Retention of Strength and Integrity Loss of strength impairs the ability to bear loads; after loss of integrity flames and gases pass through doors, walls, and ceilings. Retarded pyrolysis amounts to improved *fire resistance* and *fire endurance* as shown by *fire resistance ratings*, which are measured and expressed in units of time. A *one-hour rating,* for example, means that *the part will not collapse nor transmit flame or a high temperature, while supporting its full load, for at least one hour after the fire commences.* Chemicals incorporated into wood retard pyrolysis by minimizing the heat of combustion, or by insulating the wood from the heat of fire, through a surface layer of foam or wood's own charcoal. Insulating charcoal layers form regularly, but fire-retardants are supposed to favor charcoal formation at the expense of pyrolytic gases.

Reducing Surface Flammability and Flame Spread The improved material is less likely to ignite when exposed to heat, and when it does, the fire spreads more slowly. The spread along surfaces such as wood finish floors is measured in ft/min, for example, and expressed in *Flame Spread Classification* numbers (FSC, also called *Flame Spread Index*), which for untreated wood range from 90 to 160. *Red oak 100* and *noncombustible cement-asbestos board 0* serve as standards. Fire-retardants reduce flammability and flame spread or raise the ignition temperature by minimizing combustible gases through reduced pyrolysis, or by diluting the combustible gases, or by blanketing the surface with a noncombustible gas.

Prevention of Afterglow The glowing of charcoal after flames have ceased deserves attention because it may rekindle extinguished fires. Afterglow is prevented by covering the remaining charcoal with foam or with a blanket of inert gas, to prevent oxygen from reaching the coal surface.

Reducing Formation of Smoke At first glance, smoke appears mortally significant because when buildings burn people suffer and die from smoke inhalation; smoke causes more fatalities than heat. Killing smoke, however, usually originates in papers and rugs and furnishings which burn before wood catches fire, while the smoke of burning wood is not so critical. Less thermal degradation means less smoke, whereas reduced flammability and lower flame spread will increase smoke first due to combustible gases and tars not burning. But less burning also amounts to less heat, lower temperatures, reduced thermal degradation, less combustibles, and hence in later stages, to less smoke.

The ideal fire-retardant, which achieves all of these goals and also costs little, has not yet been invented. Some common cheap chemicals reduce surface flammability and flame spread, but they retard pyrolysis only slightly. Most notable among applied retardants are the inorganic salts diammonium and monoammonium phosphate, ammonium sulfate, borax, and zinc chloride. Commercial retardants contain several compounds; some include a resin, to bind the salts for resistance to leaching in roof shingles and other exterior products.

Few lumber dealers stock fire-retardant-treated products, but they can procure them on request from wood preservation plants and panel manufacturers. The plants press the waterborne chemical into dry lumber as described in Section 10.12 and then redry the treated wood. The treatment is either *complete* (comprising the whole cross section), or *partial* (restricted to the surface layer). Partial impregnation satisfies many needs, reducing surface flammability and flame spread; it lowers the *Flame Spread Classification* of red oak, for example, from 100 to 25 or even less. Most plants apply formulations of fire-retardants which are standardized by the American Wood Preservers' Association.

Fire-retardant plywood can be made from treated veneer, but the industry pressure-treats the panel instead, since it is high concentration in surface plies which is desirable, and sufficient amounts of the fire-retardant pass through gluelines to the inner plies. Particleboard and fiberboard manufacturers add the chemical to their binder or their furnish.

Surface coatings naturally affect the fire characteristics less than deep impregnation. The preferred coating of earlier times was water glass (sodium silicate); it formed a frothy crust when heated and thus insulated the wood from the heat of fire. Some newer coatings consist of organic substances which intumesce like water glass but have additional fire-retarding effects.

Commonly used fire-retardant salts attract moisture and raise wood's equilibrium moisture content, particularly in humid air. At relative humidities above 60% the increase amounts to several percent of moisture. Treated wood is therefore slightly weakened, having design values 10% lower than those of untreated wood. Some salts corrode metals, but corrosion inhibitors may be added to reduce corrosion to insignificant levels. Gluability and paintability remain unaffected with the right combination of glue or paint with salt. Salt crystals have however an abrasive effect on tools.

8.5. FIRE-RELATED RESTRICTIONS

Combustible material in buildings obviously should not be exposed to excessive degrees of heat. Where fire has disastrous consequences, the building codes go farther and prohibit combustible materials entirely, even if the chance of fire appears remote (Table 8.3). In some structures the codes ban wood for some purposes and allow certain wood products for other purposes, or they permit the material in particular designs.

Structural wood in multistory buildings is usually restricted to the *heavy timber construction* type with massive wood members. Beams and girders, for example, have to be at least 6 in. wide and 10 in. deep. The minima differ for other structural members. Sufficient time is supposed to pass in

Table 8.3. ***National Building Code*** **Height and Area Limitations**

Construction Type	Height (ft)	Area Limits: One Story (ft^2)	Area Limits: Multistory (ft^2)
Heavy timber	65	12,000	8000
Ordinary	45	9,000	6000
Wood frame	35	6,000	4000
Unprotected noncombustible	35	9,000	6000

fires before the wood under the layer of char loses its ability to carry the load. In the first stages of fire thick lumber has a low *heat release rate,* since it consumes much heat to heat up the moist wood, and the small amounts of evaporating gases are very diluted. For load-bearing walls some codes permit fire-retardant-treated but not untreated wood; the walls may be defined as *three-hour fire-resistive noncombustible,* for example. The fire resistance of glued laminated timber is considered equivalent to lumber the same size, with minor qualifications.

In *ordinary construction* as well as in *light-frame construction* (the category of most residential buildings), fire resistance depends on structural details like fire stops and especially on temperatures expected near furnaces, smoke pipes, chimneys and fireplaces. To avoid high temperatures, keep wood a certain minimum distance from the hot item (2 in. from any chimney) and fill the space between with loose noncombustible material to impede heat radiation and heat convection through moving air. Wood too close to hot items, for example joists above the furnace, should be separated from them by gypsum board or plaster; shields of sheet metal near hot pipes help to disperse and distribute the heat.

Completely treated wood qualifies in some building codes as *noncombustible* material. Where the codes require *low surface* flammability, for example in schools, interior finish and trim must be treated.

9

Wood, Bark, and Charcoal as Fuel

This chapter is supposed to serve readers who burn wood in fireplaces, stoves, and furnaces. One section deals with bark, a part of most firewood, showing its implications and how to handle this part of firewood best. I shall include charcoal as the pyrolysis product last to burn in fires and as the popular fuel for outdoor cooking. The reader should be familiar with Sections 8.1 to 8.3 to understand this chapter.

9.1. PRACTICALITY AND ECONOMICS OF BURNING WOOD

Fuelwood appears in two kinds: as *round fuelwood* harvested for domestic heating and cooking, and *industrial residue fuel* (sawdust, etc.), most of which is burned in wood-industry plants. Wood was man's main source of heat until coal, oil, gas, and electricity replaced it. In the United States round fuelwood consumption has dropped since 1870; during the depression, around 1930, it rose and temporarily exceeded the consumption of all other wood, then dropped again and steadily decreased to 3.5% of the total of consumed wood in 1970. In recent years, however, the high cost of oil and gas has again induced people to burn more wood.

Comparison With other Fuels

Generally fuelwood cannot compete with gas, oil, and coal. The most popular fuels—gas and oil—automatically flow through pipes into furnaces at thermostat-controlled rates, regulated by simple valves that open and close instantly. Wood can be fed automatically into furnaces too, but only with large expensive equipment such as chippers and conveyers. Wood's more difficult ignition rules out controlling the heat output by means of intermittent burning. Coal resembles wood in physical nature, but is easier to pulverize and higher in heat value (heat of combustion) even on basis of weight (Table 8.2). Anthracite coal, the preferred kind, has densities around 1.5 g/cm^3; that is three times the density of most woods, so that in heat value per unit volume anthracite exceeds wood by a factor of five.

Wood is very bulky and therefore inconvenient to handle, expensive to transport, and requires much storage space. On the other hand it is dust-free, and wood-smoke pollutes air relatively little, since it contains practically no sulfur dioxide, the most obnoxious compound in the smoke of other fuels. The little sulfur which wood does contain seldom burns to sulfur dioxide. Only 0.5 to 1% of wood weight remains as ashes; coal leaves 4 to 20%, while oil and gas burn without leaving any solid residue.

When forests are nearby, firewood costs less than oil and gas in terms of heat value, though rarely less than coal, but the potential saving is generally insufficient reason to heat with wood. However, when people have easy access to free or to inexpensive firewood, the savings may be worth their labor of collecting, preparing, drying, and handling it.

Species Differences

Since the various wood species consist essentially of the same chemical compounds, they differ little in heat value per unit of weight; all species are suitable as firewood. A ton of softwood will generate slightly more heat than a ton of temperate zone hardwood because softwood contains more lignin, hence is richer in carbon and more similar to coal. Species rich in resin and oil give up to 12% more heat than other species. Even small amounts of these extraneous substances can make a marked difference because their heat value exceeds that of major wood constituents by a factor of two. *Lightwood* is the resin-soaked wood of several softwood species, especially southern yellow pine, resin being the substance trees exude to seal their injuries. Lightwood has unusual burning qualities, deriving its name from the bright flames of its combustion, and from its use as light source and fire-lighter. It burns readily for a long time, generating much more heat than normal wood.

The heat value per unit of volume depends of course on the mass in the volume, that is, on density or specific gravity. Volumes being equal, denser woods give more heat; they also burn longer and more uniformly, since pyrolysis progresses more slowly to the center of each piece (Section 8.2). Large chunks of low-density wood burn relatively long too, but several pieces should be burned simultaneously for mutual heating. Dense firewood has transportation and storage advantages also, and in the green state it contains less moisture percentagewise than wood of low specific gravity. The high ranking of hickory firewood is justified, because—besides growing fairly straight—the hickories are in many regions of the United States the densest species.

Measurement

Firewood is measured in *cords,* a unit of volume of stacked sticks. The *standard cord* has 128 ft^3 contained in a pile 8 ft long, 4 ft high, and in the direction of stick length 4 ft deep. The *face cord* has the same 4 × 8 ft end grain front face as the standard cord, but the pile's depth is undefined, so without specified stick length the face cord represents an indeterminate measure. Solid straight roundwood in a stacked pile occupies about two thirds of the space, the other third being air. One standard cord of straight wood with the specific gravity of 0.6 contains about 3200 lb wood substance.

9.2. THE COMBUSTION PROCESS

Combustion and *burning* are other terms for *oxidation,* particularly for rapid oxidation. Since oxidation is an exothermic reaction, combustion proceeds with the evolution of noticeable heat and light, either as flames or as glow. The combination of heat and light makes fire fascinating, which is one reason for considering the combustion process here in detail, another reason being that wood is incompletely burned in many heating devices due to a shortage of information. I shall discuss incomplete combustion as such in Section 9.3 about fireplaces, in which conditions are clear and simple. For *glowing combustion* see the Charcoal Section 9.4.

Flaming Combustion

In the premixed flames of gas furnaces, gas stoves, and Bunsen burners the gas mixes with oxygen long before reaching the combustion zone. This mixture streams towards the flame at a speed exceeding the velocity of flame spread, thus keeping the flame in position so that it does not advance towards the mixing point. Pyrolytic gas of wood, coke oven gas, and natural gas all originate from organic compounds and are similar in composition; in premixed flames all three burn predominantly blue.

By contrast, gases in wood-fires emit the yellow luminous light of *diffusion flames,* in which oxygen reaches them by the irregular movement of thermally agitated molecules. Carbon particles, which heat-glow just as hot iron does, cause the yellow color; they are far too small to be seen individually and they disintegrate thermally, but persist where not enough oxygen is available for complete combustion. Luminous yellow flames indicate incomplete combustion, although above and at the side of this yellow flame the particles may and frequently do burn completely.

Diffusion flames are gases that burn with shimmering glaring light and are, as such, unsteady by definition. Degrading wood burns extremely unsteadily because pyrolytic gases vary so, changing in composition and in quantity, occasionally bursting out of new checks, and streaming from cell cavities at irregular rates. The flames are long where gases puff out in large amounts, they blanket the wood when gases evolve evenly over a large surface. The dancing flames of various colors consume combustible gas on one spot, then jump to adjoining areas of accumulation, traveling back and forth on the vehicle of fuel supply.

Smoke, Creosote, and Soot

Wood's main chemical elements—carbon, hydrogen, and oxygen—burn to the invisible gases water vapor and carbon dioxide. Consequently wood should burn without smoke, whereas in reality many wood fires smoke considerably, due to solid particles and droplets in the fire effluents. The carbon particles of diffusion flames contribute to smoke when cooled to the point where glowing stops. Ashes swept up by the air current represent other smoke solids, while tar, water, and organic condensates form the droplets.

Smoke color depends on the size of the particles and droplets. Invisibly small sizes, smaller than the wavelength of light, make the smoke white or bluish as in the smoke of cigarettes. Smoldering fires, which burn with a minimum of flames and a maximum of smoke, typically make a white smoke.

The kind of fuel determines the color of smoke made up of large particles and large droplets: carbon causes black smoke, tar and other organic products brown smoke. The color of the water fraction again depends on the size of the droplets, as we can observe from the minute droplets within light fog and white clouds, and from the large rain drops coming out of dark clouds.

Organic vapors in fire effluents condense on cold chimney walls as *creosote,* a brown or black stenchy liquid which resembles the wood preservative *coal tar creosote* (Section 10.10). Chimney creosote also forms from precipitated tars. Fresh creosote is still fluid, being condensed acids plus water, but evaporation leaves it sticky, tar-like. Long exposure to heat pyrolyzes chimney creosote to a flaky layer of carbon known as *soot.* Carbon particles tend to aggregate and adhere to chimney walls, augmenting the soot.

Resinous wood species like pines, spruces, and Douglas fir have the reputation of generating excessive soot and creosote, possibly because they burst into flames accompanied by much smoke in the early stage of burning. The factors of moisture, low temperature, and lack of oxygen are definitely much greater in creosote and soot production than that of resin. In hot fires with sufficient oxygen the substances burn completely, leaving nothing to be deposited, while a smoldering fire of wet wood produces excessive soot and much fluid creosote. Weak fires are fuel-inefficient.

Fresh creosote is annoying when it leaks out of stove-pipe joints and from pipes corroded by condensed pyrolytic acids. Creosote and soot have to be cleaned out periodically, above all to prevent chimney fires, but also to improve heat transfer from the fire effluents through the chimney walls into the house, and to smooth out the flow of effluents.

Effect of Moisture Content on Fuel Value

Moisture impairs the heat value of wood more than any other variable. Theoretically, wood's heat of combustion can evaporate 720% moisture (compare Table 8.2), but vapor in fact interferes with the combustion process and prevents pyrolytic gases from burning. The amount of heat actually generated in furnaces, called *fuel value* in Fig. 9.1, is therefore lower than the calculated *heat value* at that moisture content.

The *dry weight basis* curve of Fig. 9.1 shows the effect of moisture on the fuel value of a given quantity of wood. Compare for example the fuel value of a pile at 25% and at 100% moisture content: their ratio is about *7400:5000*—the dry pile gives 1.5 times more heat than the wet pile.

The *wet weight basis* curve serves wood-burners who buy by weight. When comparing for example one ton of firewood at 25% with a ton at 100% moisture content, the ratio of the two fuel values is about *6000:2500*, in this case the dry wood giving 2.4 times more heat than the wet wood. On wet weight basis, moisture affects fuel value so much because the wet ton contains much less wood substance (0.5 ton) than the dry ton (0.8 ton).

Figure 9.1 applies to good furnaces only. Most stoves are less efficient and burn the wood less completely, particularly at high moisture contents, so that wood sustains a fire only if its moisture content is far below 200%.

Owners of wet firewood have the option of burning wet, that is, drying wood within the firebox, or drying it before burning, for example outside the firebox but within its zone of warmth. Burning wet involves less work but wastes energy. *Dry, then burn* is more energy-efficient than *dry and burn.*

9.3 BURNING WOOD IN FIREPLACES

Open fire appeals to young and old alike. Man enjoys the play of flames around the crackling wood, from first ignition to the final glowing charcoals.

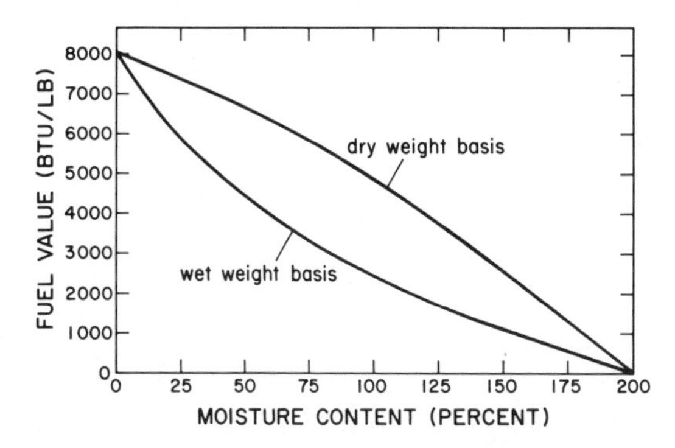

Figure 9.1. Relation of fuel value and moisture content in good furnaces, not including the latent heat of water vapor.

Open fire radiates heat immediately, bringing quick comfort to a cold room and dispelling morning and evening chills in spring and fall. However, a traditional open-burning fireplace is surprisingly inefficient. To gain the same amount of heat as from a good furnace, one must burn on the average ten times more wood.

Good heating equipment burns the fuel completely and transfers the generated heat out into the house. To this end furnaces and stoves consist essentially of two parts: a firebox for combustion and a structure for transferring the heat out. The fireplace ideally does both—burning the fuel and transferring the heat; but is quite deficient at each task. In a fireplace the fuel burns incompletely and most of the heat generated escapes through the chimney. Additionally, fireplaces waste heat by drawing excessive amounts of warm air from the house. The deficiencies are discussed here first before dealing with fireplace fuel.

Air By-Pass and Air Supply

Good burning devices channel all entering air into the fire and draw no more air than needed for combustion. Fireplaces draw plenty of air from the warm room directly into the chimney, the air flowing mainly through an area of least resistance, the space between fire and upper front edge of the fireplace (Fig. 9.2). The chimney-draft pulls equal amounts of cold air through joints at doors and windows into the house, so that in cold weather the fire actually cools the warm house instead of heating it.

The damper (Fig. 9.2) controls airflow, but it cannot reduce the percentage of air bypassing the fire, and it must be relatively wide open to prevent smoke from entering the house. *Fireplace doors* with air-intakes below grate level

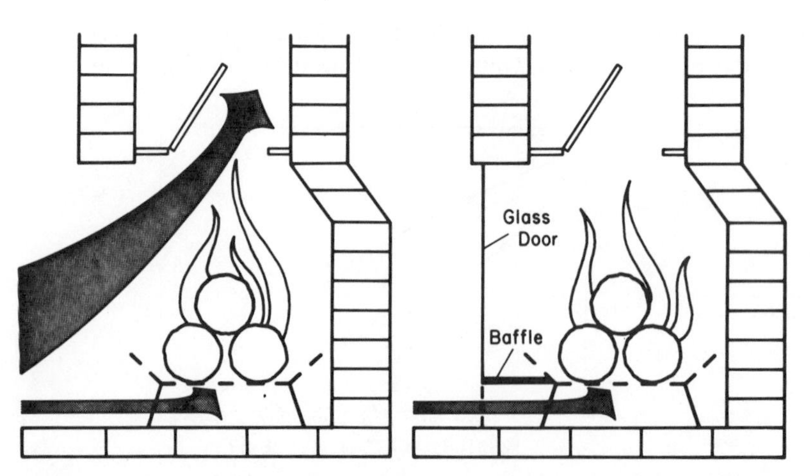

Figure 9.2. Flow of air in fireplaces. (Left) Traditional open-burning type. (Right) With glass door and baffle.

(Fig. 9.2) can throttle this by-pass and virtually prevent it, if supplemented by a baffle extending from grate to door as indicated in the Figure. Some modern *closed* fireplaces draw air through pipes from outside the house to avoid any heating loss through the draft intake.

Ideally all entering air passes through the degrading wood, mixing with pyrolytic gases which glowing charcoal ignites to produce a low fire of short flames. Excessive amounts of air drawn into the fire zone from the side dilute the pyrolytic gases and chill them below ignition temperature, contributing to incomplete combustion which is discussed next. Remember, blowing toward glowing charred wood speeds up a smoldering fire, but blowing into the flames stops combustion.

Incomplete Combustion

Flames and glowing coals radiate heat. In furnaces and stoves the radiation heats the firebox walls, which in turn keep the fire hot. Open fire radiates the heat out and for this reason cannot become very hot; its temperature remains too low for spontaneous ignition of gases (Chapter 8.3), which escape without burning except where licked and ignited by flames. Due to low temperatures, single large chunks of wood do not burn in fireplaces, instead several pieces which heat each other have to be burned simultaneously, just far enough apart to give oxygen access to their glowing charred surfaces. It has been suggested that logs be stacked rackwise above each other in the form of a *wall,* making a large surface to radiate heat out into the living room. Whether the firelog *wall* will provide sufficient mutual heating depends on conditions.

Ashes under the grate restrict airflow to the grate and reflect heat up into the fire, in this way slowing the fire without lowering its temperature. *Ash banks* control the fire; a *well-banked* fire stays alive for many hours. The charcoal in heat-insulating ashes slowly burns through the whole night and rekindles fresh fuel the following day.

Heat Transfer in Modified Fireplaces

In the last decade numerous devices have been marketed intending to make fireplaces more efficient, some by transferring heat out of the fireplaces, others by reducing the air bypass. Most notable in the heat transfer category are tube grates, which move heat out of the zone of fire effluents, but may also contribute to incomplete combustion by cooling pyrolytic gases in the fire zone.

Fireplace glass doors, installed to reduce air bypass, create a sort of firebox. Heat radiated by the fire is absorbed by the glass which heats up and keeps the fire on the higher level of temperature needed for complete combustion. On the other hand, the glass restricts primary radiation (by absorbing the radiated heat) and traps heat in the fireplace, so that ultimately much escapes through the chimney. The doors transfer some heat out: through air that heats up and rises at the room-side surface, and through secondary radiation by the hot glass, but the transfer cannot be as intense as the primary radiation of open fireplaces. The doors' net effect on fireplace efficiency is uncertain; in case of small and dying fires they help by reducing air by-pass, but in strong fires the restricted primary radiation may cancel the gain.

Fire radiates heat to the firebox lining. In stoves the lining conducts absorbed heat to the outer surface in the living room, whereas in fireplaces most heat remains stored in the lining. One improvement is to circulate room air through ducts of heat exchangers in the lining, or around steel fireboxes within the fireplace. Still better in this regard is the fireplace type known as the *Franklin stove.*

Fireplace Fuel

Wood of high specific gravity is preferred in fireplaces for the same reasons it is in other heating equipment. Softwoods burst into high flames and burn up readily. The highest priced firewood species seem to be paper birch and gray birch, their white bark being a most important feature for people who rarely light fires but have a showcase hearth. Birch is also known to burn well, as explained in Sections 8.3 and 9.5. Due to the unfavorable conditions in fireplaces, firewood must be drier than wood for furnaces and stoves; it does not burn if the moisture content exceeds about 80%.

Firewood gives off smells until the chimney has drawn the odors away. Most notable in this respect is piñon pine. Pleasant aromas allegedly emerge also from wood of fruit trees, such as apple and cherry, and the nut trees hickory, pecan, and beech, whose smoke is supposed to recall the fruit fragrance.

Several woods *pop* and throw *sparks.* The phenomenon has been reported for hemlock, larches, spruces, butternut, chestnut, locust, white birch, and hickory. It has never been thoroughly investigated, but may be due to pockets of steam, vaporized volatile portions of resin, and pyrolytic gases which build up under the wood surface when the temperature rapidly rises. Surface layers supposedly plug the pockets until pressure rises to the explosion point and shoots the burning plug out. Moisture favors the popping when filling steam pockets, but works against it by retarding heating. The sparks fly perpendicular to the fiber direction from lateral wood surfaces; therefore place sparkthrowing logs with their end grain toward the fireplace opening. Only bare wood surfaces seem to throw sparks, bark and thick char do not.

Some people prefer *fireplace logs* or *firelogs* made from compressed sawdust and other small-size wood residues. The logs are readily available as handy clean packages of uniform properties, always dry and ready to burn. One kind consists of sawdust pressed together under heat to achieve self-bonding. Having specific gravities near 0.8 the logs are as difficult to ignite as wood of same density, but after ignition they burn faster than solid wood because many fibers run perpendicular to the log axis and conduct the heat relatively well. The logs also break up into short sections. In some firelogs the wood particles are bonded with plenty of wax, a petroleum refuse which together with the wrapping paper facilitates ignition, raises the heat value, and burns steadily for a long time.

9.4 CHARCOAL

In the early stages of wood fires, pyrolytic gases blanket the charred wood surface and impede its oxidation. When thermal degradation later diminishes, plenty of oxygen molecules reach the hot coal and react with carbon causing the charcoal to burn and glow. Whether the glow comes from combustion as such, or is a result of the high temperature (heat-glowing), is an interesting question. Since the temperature of spontaneous charcoal ignition (observed between 300 and 480°F) is much lower than that of incipient heat-glowing (885°F), and the heat originates from combustion, glowing **does** indicate burning.

Charcoal burns in two steps, first to form carbon monoxide gas, which combines in the second step with oxygen to form carbon dioxide. Some hydrogen remaining in the degrading wood escapes at the high temperatures and burns together with the carbon monoxide. Neither hydrogen nor carbon monoxide can form carbon particles; both therefore burn with barely visible nonluminous diffusion flames.

As a fuel, charcoal has several advantages over wood: it burns without smoke at constant rates, because combustion is in the first step a surface reaction; the fire is hotter and generates nearly 1.7 times more heat per unit of weight (Table 8.2). On the other hand charcoal's specific gravity is only two thirds that of the original wood, and thermal decomposition in the production of charcoal takes away at least one half of wood's heat value. Hence, in heat value per unit of volume charcoal ranks far behind wood. But in the form of briquettes compressed to a specific gravity near 1.0, the heat value per unit of volume by far exceeds that of wood.

Fresh charcoal as it leaves retorts is highly reactive and ignites easily. The briquettes do not retain this reactivity; like natural charcoal, they release no combustible gases at first, and are therefore much more difficult to ignite than wood. To facilitate ignition, producers incorporate flammable substances or consumers spray the coals with lighter fluid a few minutes before lighting. The coals readily absorb fluid, since the briquettes are still porous (charcoal cell wall substance, having the specific gravity 1.5 of wood cell walls, comprises only two thirds of briquette volume). Charcoal ignition depends a great deal on the storage conditions. The coal remains hygroscopic even though its equilibrium moisture contents are lower than those of wood. Stored in cool humid corners the coal will absorb moisture, which has to be boiled out before the temperature can rise to combustion level.

9.5. BARK AS FUEL

Since most wood is burned within its bark, some related properties of this part of the tree deserve attention. The properties make sense when considered in context with the tree, and are explained here first to form the basis of bark's burning characteristics.

Bark as Tree Tissue

As illustrated earlier (Section 1.3), the cambial layer produces wood cells to the inside and bark cells to the outside. The ratio of bark growth to wood growth differs from species to species; oaks, for example, grow more bark than beech. The proportion of bark in the tree trunk volume falls behind as the tree grows in diameter, because bark lets up in growth and sloughs off at the trunk surface. In young white oak 5 in. in diameter almost one third of the trunk is bark, whereas in the mature trunk the bark diminishes to about 10%; in old beech trunks this proportion drops to 5%.

Bark consists of two main layers: young *inner* bark close to the cambium and old *outer* bark at the very surface of the trunk. Inner bark corresponds to sapwood and is of similar age; in living trees inner bark contains living cells whose sap in some tree species amounts to a moisture content of 300%. Outer bark, corresponding to heartwood, consists of dead cells only. A thin membrane separating inner and outer bark retards migration of moisture to the outside, both to protect the three's vital cambial area from drying out and to keep outer bark dry. The outer bark contains little or no free water; at the very surface its moisture content approaches equilibrium with the humidity of the atmosphere, rarely rising above 20%.

In living dense hardwood trees the moisture content of bark as a whole averages 50 to 90%. The light bark of light wood species has plenty of space for free water and contains at least 90% moisture. In standing dead trees, bark naturally dries first and retards drying of the wood somewhat more than a wood layer of equal thickness would, but in short stem sections of stacked firewood, which dry mainly through their end faces, bark permeability does not matter.

Bark, particularly the dry outer layer, protects the tree's cambial area from extreme temperatures, insulating the living cells from daily waves of heat and cold by equalizing temperature fluctuations. Bark conducts less heat then wood of same moisture content, but this difference is small.

Adhesion of Bark

Drying strengthens the cambial bond between bark and wood, just as it strengthens wood. But when the unlignified soft cambial layer decays during the drying process, bark falls off. In living trees the cambium resists decay because decay bacteria and fungi need oxygen, for which the sap-soaked cambial cells have no space. In felled trees, air penetrates the drying cambium, but then decay organisms may soon fall short of moisture. Thus fungi and bacteria grow and destroy the cambium within a certain crucial moisture range only.

During drying of firewood, the cambium inevitably reaches the crucial range. Whether decay organisms then consume it depends on temperature and how long the range prevails. In frozen timber as well as in timber above natural temperatures, bacteria and fungi become dormant and do not cause decay, no matter how long the crucial range persists, whereas in the favorable temperature range between 70 and 90°F the cambium may decay in a few days. Below 50°F it endures, holding the bark for months. In short pieces stacked in open piles immediately after tree-felling, the cambium dries during pleasant summer weather too quickly for decay, at least in sticks of small diameter and in split logs. In winter when drying is slow the crucial moisture range may prevail for months, but then low temperatures retard or arrest the decay process.

Ignition and Combustion

The bark of green tree sections burns readily as long as dry outer layers decompose, but as soon as the heat wave reaches wet inner bark and converts free water into steam, the fire lets up. Outer bark that is partly detached from the trunk—as happens naturally in shagbark hickory—is especially dry and burns easily. Detached bark also ignites readily because it conducts no heat to layers underneath, so there are no vapors effusing out of deeper layers to dilute combustible gases of the quickly heated surface. Easiest to ignite of all parts of any tree are the papery, resinous bark curls of birch.

In fire, attached bark first acts to insulate the wood from heat until it has burned away. Since bark consists of almost the same chemical elements as wood, its heat value on weight basis is nearly identical at the same moisture content. In some species bark ranks above wood in heat value, in the majority wood has more energy, although the difference amounts only to a few percentage points. Lower heat values of bark are explained by its higher ash content, which on the average in North American species is 4%, and reaches 10% in some tree species. In most species the higher specific gravity of bark makes up for the ash, so that its heat value per unit volume equals that of wood. Attached bark contributes to the fire essentially like a wood layer of the same thickness.

10

Degrading Organisms, Durability, and Preservation of Wood

Most plants live tediously on inorganic material, absorbing water from the soil and carbon dioxide from air to synthesize organic matter. All trees do this, and their main product is wood. Other plants and all animals obtain nourishment from relatively rich living or dead tissues of still other plants and animals. Amidst plenty they thrive and multiply, until supply and demand of the organic food balance out.

Under such conditions organic matter tends to be eaten up almost as fast as it becomes available. Wood makes an obvious exception, accumulating in trees in larger quantities than are consumed, until after many decades the forest reaches climax stage. Wood accumulates because most living organisms lack the ability to digest this tough lignin-hardened substance. Man's inability to eat wood is typical: he may chew and swallow small pieces but his organs cannot break them down chemically into water-soluble compounds, transportable in the bloodstream to places of oxidation in muscles and other parts of the body.

Some plants and animals, however, can digest wood and live on it; they specialize just on this plentiful material. Other animals use wood for shelter and work themselves inside, utilizing it as food at the same time if they can. Organisms that feed on wood and live in wood are the natural wood degraders, shortening wood's lifetime, though only under certain conditions. This chapter explains these conditions and how the degraders live, so that the reader can use wood wisely. I shall deal with degrading organisms in the sequence of fungi, bacteria, insects, and marine borers. Then comes a discussion of the closely related durability of wood, and finally wood preservation, or poisoning against its degraders.

10.1. PHYSIOLOGY AND REQUIREMENTS OF FUNGI

Fungi are plants with no flowers, leaves or green color, which reproduce by means of spores. The lack of green color indicates absence of chlorophyll, meaning no photosynthesis or need for light, and therefore the ability to live deep in a substrate such as wood. Fungi grow in the form of thin filaments, called *hyphae* (singular *hypha*, from a Greek word for web), primarily within the cavities of wood cells (Fig. 10.1). The threadlike hyphae are so thin that it takes some effort to spot them even under a microscope.

Wood-inhabiting fungi absorb their nutrients in aqueous solution (as do all other plants) hence they need *free* wood moisture available to them, at fiber saturation. As fluctuations of temperature and humidity in the surrounding air cause uneven moisture distribution, some free water may accumulate locally even when the average moisture content of the piece is below fiber saturation. Fungi have in addition the ability to utilize some *bound* cell wall moisture, just as common plants can draw water out of seemingly dry soil. Wood is therefore safe from fungal attack only if the average moisture content stays below 20%, in the *dry* state as defined in the *American Softwood Lumber Standard* (Section 5.1).

Since many wood products pass through wetting-drying cycles, the question arises whether fungi survive in dry wood—they do, differing in this respect from higher plants for which drying means death. The fungi's survival period has a limit, though; it varies from a few weeks up to many years of dryness, depending on the fungus species.

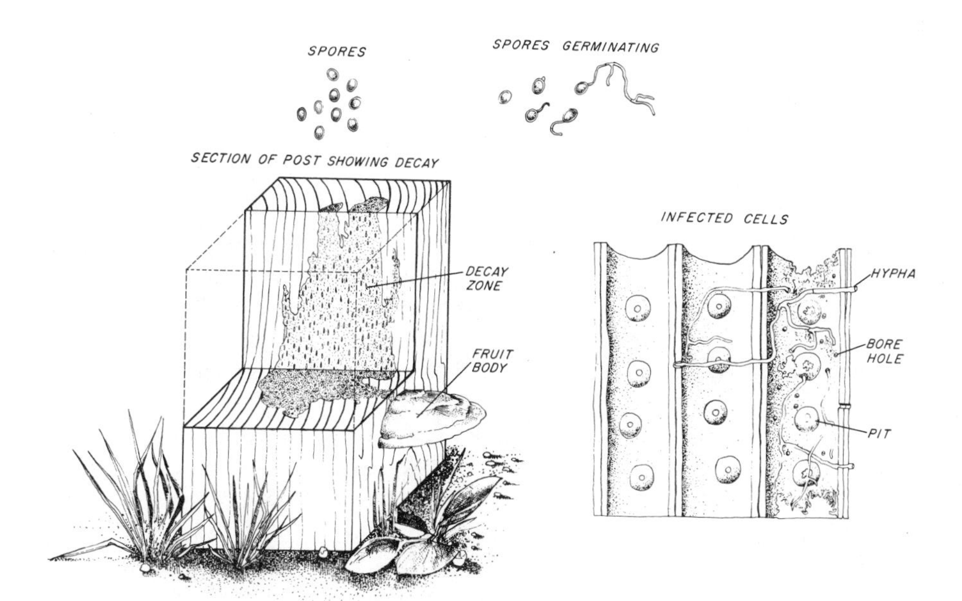

Figure 10.1 Cycle of decay. (From Scheffer and Verrall, 1973.)

Animals and plants that live on organic matter need oxygen for respiration. For fungal growth also wood must contain air, it cannot be in the *water saturated* state. Wood under water or in mud is therefore fairly safe from fungal attack.

Decay begins in many living tree trunks in heartwood and not in sapwood, because heartwood contains air which entered through broken branches. In sapwood the trees themselves drown fungi, and living parenchyma cells compete for dissolved oxygen. The fact that fungi need water and oxygen is the reason why fence posts decay first at the ground level, not underground where oxygen is short, and not in the dry part above ground.

The need for oxygen determines the air space required and the upper moisture limit. The limit—expressed in percent moisture content—is lower for dense wood than for light wood because dense wood has less cell cavity space, less space for water, therefore a lower moisture content in the water saturated state (Section 3.4), and a lower moisture content when the wood contains the minimum of oxygen required by the fungi. Optimum moisture lies only a little above fiber saturation; water is so dense that small volumes of it satisfy the fungi's need, whereas the air space must be quite large for sufficient diffusion of oxygen from the wood surface to deep layers. Additionally the carbon dioxide that fungi respire must diffuse out to prevent them from suffocating in their own waste gas.

The temperature optimum for growth of most wood-degrading fungi lies in the same range as the optimum for other plants, between 70 and 90°F depending on species. Outside this range fungal activity slows down. Frost brings all plant growth to a halt because frozen water can neither serve as medium for transport nor participate in metabolism, but frost as a rule does not kill fungi, otherwise they could survive cold winters only deep underground. Heat above 100°F is lethal for most fungi if the exposure lasts long; how long depends on the kind of fungus, the resistance being greater in their dry-dormant state than in their moist-active state. Kiln drying kills practically all fungi. High temperature drying above 100°C (212°F) destroys even spores, but this has little significance because fungal spores abound everywhere.

Fungal spores are microscopically small; wind, water, and animals spread them in billions all over the earth onto every object. Fungi also spread by growth into adjoining wood and by tissue fragmentation when infected pieces of wood are carried away to other places. Fungi in these tissue fragments, being already well established in the wood substrate, can resist adverse conditions better than tender germinating spores and are a certain cause of new infestation where conditions permit fungal growth.

When no fungi appear on wood after some time the reason is not absence of spores and fungal fragments, but simply an unfavorable environment. The most favorable conditions exist in the ground close to ground surface. There the soil harbors plenty of spores, possibly also fungal fragments, some of which are likely to prefer the particular wood species. Soil water keeps the wood moist, plenty of oxygen is accessible, and temperatures include the favorable range.

Fungal growth aboveground depends on precipitation, humidity, and temperature. For the United States, researchers have calculated a corresponding *climate index* based on days in the month with 0.01 in. or more of precipitation and on the mean monthly temperature. Figure 10.2, drafted on this basis, indicates favorable conditions for fungi in warm moist areas and unfavorable conditions where either temperature or precipitation or both fall short. The chart naturally also reflects conditions in the upper layers of the ground.

10.2. MOLDS AND STAINS

Molds are fungi that have fuzzy, fluffy, or powdery surface growths. Species that grow on wood lack fuzz except in extremely humid air; they penetrate the wood but are macroscopically in evidence only at the surface. Masses of spores on softwoods and tissue pigmentation on hardwoods cause spotty surface discolorations in various tones. By contrast, *stain* fungi discolor the wood in depth; their hyphae contain or release a pigment which when seen en masse through surface cells gives stained wood an unattractive appearance. The most prevalent

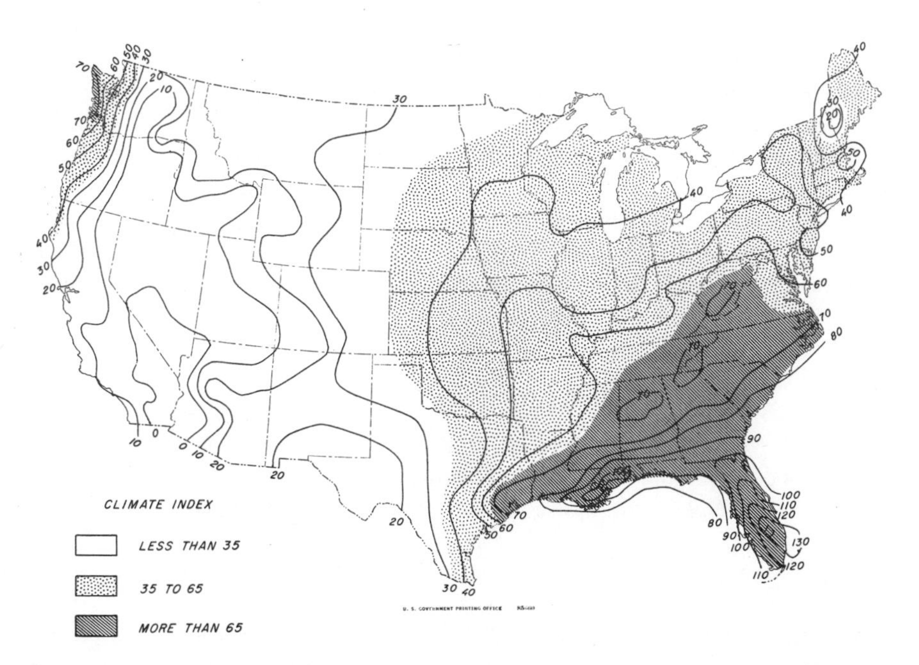

Figure 10.2. Levels of decay potential for wood in aboveground service, exposed to weather, based on a climate index derived from standard temperature and rainfall data. Darker areas are more suitable for decay. (From Scheffer and Verrall, 1973.)

group, *blue stain*, makes the material look bluish or bluish-black. Other stains cause different discolorations.

Mold and stain fungi live on readily available sugars and starch in cell cavities, as well as on parenchymatous tissue concentrated mainly in the rays. The stainers prefer to advance along rays, causing radial streaks of discoloration. When growing from cell to cell the hyphae of molders and stainers break mechanically through the pits, making the tissue more permeable so that infested wood absorbs water, wood preservatives, glue, and finishes several times faster than sound wood. Since absorbed water and permeability favor decay fungi, molders and stainers prepare wood for decay.

Stains and molds grow in sapwood of most (but not all) wood species. Both groups avoid heartwood because it lacks readily digestible carbohydrates. The wood industry therefore prefers the name *sap stain* fungus over *stain* fungus.

In advanced stages of infestation, stain and mold fungi may attack the wood substance itself, weakening the tissue especially in impact bending and toughness. Heavily stained wood should not be used where strength is very important because discoloration signifies that conditions were favorable for decay and that decay infection may already have occurred. Stain and mold may mask decay; in stain-infested wood decay remains invisible for a long time.

So far these fungi appear to be relatively harmless. They grow however under a wider range of conditions than other wood-degrading fungi, at moisture contents as low as 20%, and so cause more complaints in construction. Mold and stain develop quickly since they live on rich, readily available nourishment. Under favorable conditions the two can spoil whole shipments of lumber within a single week, in humid weather even degrading lumber while it air-dries. Stained wood is unsuitable for many purposes, whereas moldy material just requires shallow planing, sanding, or simply brushing to restore its beauty. Stain does, mold does not lower the grade of lumber.

10.3. DECAY FUNGI

Species in this category have the ability to decompose wood throughout. Decay fungi break the very wood substance down, dissolving and absorbing it almost completely. They cause what is commonly known as *wood rot*.

Life Cycle and Growth

Life of an individual decay fungus begins with a spore, which at the proper temperature, on suitably moist wood of the right kind, germinates and grows hyphae into the tissue (Fig. 10.1). Hyphae, when advancing from cell to cell, first choose easy paths through pits and rays, rendering the material permeable and absorptive, as do stains and molds. In a later more aggressive stage, the hyphae secrete enzymes which etch *bore holes* into the cell walls for them to grow through. Further secretions widen the holes, corrode whole walls, and convert wood substance into water-soluble components to be absorbed.

As hyphae spread and ramify, some grow together in vine- and bandlike strands to transport dissolved compounds and conduct water from wet to water-deficient zones (Fig. 10.3). Some species can send strands into surrounding wet soil to gather water. Their apparent ability to decay dry wood has gotten them classified inaccurately as *dry rot fungus*. The moisture minimum of decay fungi is higher at first than that of molders and stainers, but when established they utilize moisture from decomposed wood and may need less.

The hyphae grow in cell cavities and bore holes, later forming strands between decomposed and shrivelled wood tissue. Some mycelium, that is, the mass of hyphae and strands, spreads on the surface as patches, especially in damp stagnant air where it does not dry out. The mycelium of most species of fungi lacks color and is transparent, but thick strands and patches appear whitish and hide the foul-smelling wood with a soft, sordid skin (Fig. 10.3).

Mature mycelium develops spore-bearing fruit bodies from interwoven hyphae. To make it likely that spores will spread to other wood, the fruit bodies grow only at exposed wood surfaces. The fungi use light as indicator of exposure, although very intense light will prevent fruiting. Some fruit bodies have a stalk-and-cap structure like that of mushrooms (some mushrooms in fact grow on wood), but most species of wood decay fungi have

Figure 10.3. Water-conducting strands of brown rot fungus terminating in a mycelial *fan*. (From Scheffer and Verrall, 1973.)

bodies of the bract or bracket type shown in Fig. 10.1. Each fruit body releases millions of spores, enough to infect every suitable substrate and repeat the fungus' life cycle.

Effect on Wood

Decay is a type of slow fire. The fungus oxidizes degraded absorbed wood compounds to carbon dioxide and water, reversing the tree's process of photosynthesis. Carbon dioxide as well as some vapor diffuse into the air and the decaying wood loses weight. Wood breakdown involves loss of strength, whereas, surprisingly, its stiffness remains nearly unchanged until the final stage of decay. The first visible symptoms of decay—loss of luster and slight color change—appear only after significant weakening.

The fungi affect strain at failure early and substantially, so that slightly decayed sticks snap when bent only a little, and dynamic bending strength as well as work to maximum load drop sharply to one half when the weight loss is still only barely detectable. The early loss in strain at failure is due to the splitting of linkages in chains of carbohydrates, a disintegration which indeed can occur without weight loss. Low strain at failure is the basis of the *pick test* for early decay: when poking into decaying wet wood with a pointed tool you cannot lift out long splinters, instead the wood breaks abruptly like brash dry wood in sound condition does (Fig. 10.4). Decaying wood also fails in bending with a short or brash fracture, as opposed to the long-fibered or splintery fracture of sound wood.

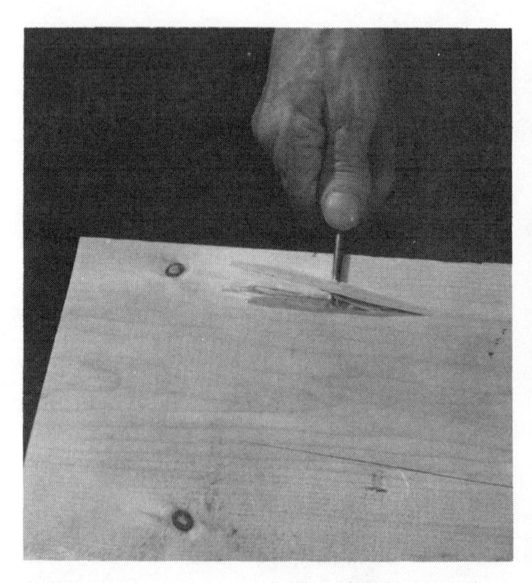
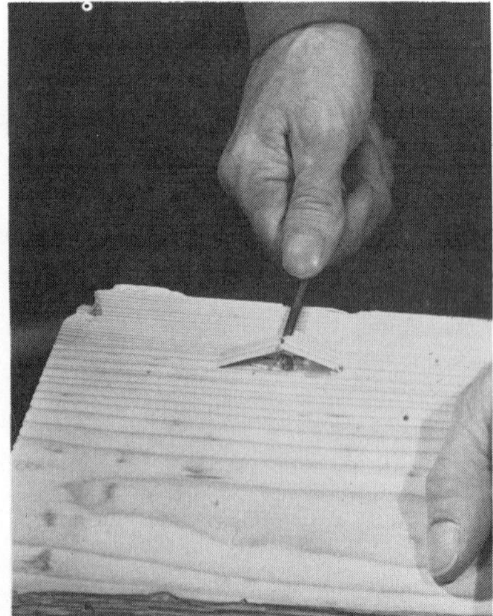

Figure 10.4. *Pick test* on sound wood (left) and on infected wood (right). (From Scheffer and Verrall, 1973.)

Mechanism and Rate of Decay

The secreted enzymes catalyze wood decomposition by linking with functional groups of wood constituents, such as with the glucose residues of cellulose molecules. This linking changes the characteristics of the residue and breaks it out of the cell wall. Characteristically the enzyme is not consumed in the reaction, it simply drops off and links up with another functional group, working in this way at high efficiency in small quantities, though slowly. It is enzymes that determine fungal growth and the course of decay.

Moisture participates in this process in three phases: (1) the enzymes diffuse in liquid water from the hypha to the cell wall surface; (2) water takes part in the *hydrolytic* break-up itself; and (3) the fungus absorbs the residues in aqueous solution.

Decomposition is a slow, tedious process. Even small pieces of nonresistant species resist decay under favorable conditions at least two months; in thick durable heartwood, at marginal temperatures and moisture contents the process goes on for decades. Landowners notice the slowness of decay when waiting for a tree stump to rot out of their way (under a cover of soil the stump rots a little faster).

Types of Decay

The fungal species secrete different enzymes. *Brown rotters*, named after the discoloration they cause, secrete one for cellulose, raising the percentage of brown lignin in the residue. This residual altered lignin maintains cell shape until the cells are deprived of their cellulosic framework and collapse with large contraction in all directions, not only across the grain. Due to restraint by less decayed deep zones, the contraction involves checks along and perpendicular to the grain, along with formation of characteristic cubical remnants as on charred lumber (compare Figs. 8.2a and 10.3). The color of brown-rotted wood and its crackedness create the impression of dryness, another fact which contributed to the misnomer *dry rot.*

White rot fungi live upon all major wood constituents and destroy them all at the same rate, leaving behind whitish decayed wood within black zone-lines formed by masses of hyphae. The bleaching results from destruction of colored matter, not from removal of brown lignin as has been suggested. White rotters attack the cell as a whole, beginning from the cell cavity and working towards the *middle lamella* between cells; they cause excessive contraction and collapse only after decay is severe.

Heartrot in the heartwood of standing tress is brown or white and is caused by species of fungi that specialize on standing trees, mostly of a particular species. Due to this specialization, heartrot fungi generally cease to grow in lumber, while conversely a standing tree is safe from lumber-decay fungi. In some softwood trees certain fungi cause pocket-type rot, in which comparatively firm wood contains scattered pockets of advanced decay. The wood is valued for its exotic appearance in *pecky* cypress and *white-pocket* Douglas fir paneling.

10.4. SOFT ROT FUNGI

Soft rot derives the name from the soft wood surface produced, which is easily scraped off from the sharply distinct sound wood beneath. Soft rot looks very much like brown rot, with profuse fine cracking and fissuring caused by restraint of contraction from the sound wood below (Fig. 10.5).

Soft rotters do not form fruit bodies as the decayers do; they are closely related to stainers and molders. Under proper conditions, in fact, stain and mold producers cause soft rot. The difference lies in their level of aggressiveness and in time; soft rot is a late aggressive stage of stain and mold. The fungi in weathering wood surfaces (Section 14.1) seem essentially to be the soft rot kind, only their action is too slow and their effect too indistinct to be called *rot*. Unpainted, severely weathered lumber eventually gains the typical appearance of soft rot.

The best known examples of soft rotted wood are submerged timbers in rivers and slats of cooling towers, over which water is cascading in a continuous stream. Soft rot damage has been observed on many other wood products, including utility poles and railroad ties, but it is greatest in thin pieces.

Researchers do not know for sure why soft rot tends to remain restricted to shallow surface layers and why it succeeds where decay does not. It seems

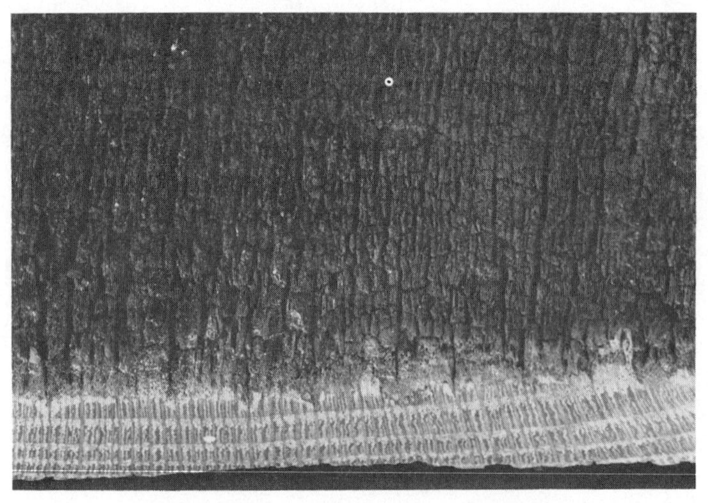

Figure 10.5. Soft rot infected oak board.

to require less oxygen than decay and thrives where decay fungi cannot breathe, but in deep layers even soft rotters have too little oxygen, a view evidenced by soft rot which over long periods of time—when sufficient oxygen finally penetrates—does grow into deep layers. In slats of cooling towers and submerged timber, dissolved air satisfies the fungi's need for oxygen at the wood surface but not in deep layers. Soft rotters can also tolerate drier conditions than decay; they simply demand less, a fact explainable by their lower development in the evolution of plants.

10.5. WOOD-DEGRADING BACTERIA

Like fungi, most bacteria depend on tissue of other organisms, do not need light, and have the ability to flourish deep inside the substrate. These single-celled organisms thrive in wet heartwood of some living trees, in stems of log ponds, sometimes in packs of green lumber, and in wet piles of sawdust (as signified by the sour smells exuding from infested materials). Bacteria obviously need lots of moisture, at least as much as fungi. Of oxygen they demand less; the anaerobic kinds abound even in complete absence of oxygen. Under these conditions we may expect bacterial activity in all living trees and in all wet lumber. Yet bacteria play only a minor role as wood degraders, which is surprising in view of the speed with which they spoil food and cause all sorts of diseases. Why in fact they do not ruin our forests is due to acidity; wood pH is too low for their liking.

The relatively rare wood-inhabiting bacteria live first on food reserves stored in the soft parenchyma cells of rays and on easily digestible membranes of pits. Their attack makes wood permeable and absorptive for all sorts of liquids including water, thus paving the road for decay.

Infestation has no effect on wood strength in the early stages. Only after long periods of time do secreted bacterial enzymes break up lignified wood substance and weaken the material. Wood pilings in fresh water or in muddy ground under building foundations may finally deteriorate from bacterial action. In wet piles of sawdust and of other small wood, where the tiny organisms can metabolize much faster, decomposition progresses rapidly and generates so much heat that the temperatures rise up to 180°F, much higher than the temperatures tolerated by fungi.

10.6 COMMON FEATURES OF WOOD-DEGRADING INSECTS

The number of known species of insects approaches one million. Though only a small percentage of them live in wood, the wood-degrading species are still legion and impress one with their tremendous variety. They differ greatly in appearance, in life habits, in requirements, in damage, and in vulnerability.

Life Cycle and Requirements

Wood-degrading insects reproduce by means of tiny eggs, usually deposited by the adult female into vessels, checks, or holes which she punctures in the wood surface. From the eggs hatch larvae (*worms*), which grow in the wood to adult form, some after passing through a pupal stage. In many species the adults *bore* out of the wood substrate to mate and to lay eggs again. This life cycle usually lasts many months, in some species several years, and in dry wood longer as a rule than in moist wood for the same insect species. Typically insects live by far the longest part of their life as larvae and they damage the wood mainly in this stage. Social insects like termites and carpenter ants make the exception; they live most of their life as working adults, pampering the larvae as benefits an organized society.

The insects prefer wood moisture contents at around fiber saturation, where the substrate contains sufficient oxygen, is relatively soft for chewing into, and the humidity within tunnels prevents the young thin-skinned larvae from drying out. Sufficient moisture seems to be the main reason why many species select standing, dying, and recently felled trees. Some kinds however specialize in dry wood; many start in green timber and complete their development one year or so later in dry lumber.

Some species feed on the very woodiest substance and relish even heartwood, provided it contains no poisonous substances, but most species depend on the starches, sugars, and other readily digestible foodstuffs in sapwood. One group derives nourishment from mold-type fungi which the females implant and cultivate in holes they make in moist wood, causing stains which usually disfigure the material more than the holes do. Several species are scavengers, which concentrate on soft rotten wood.

Typically an insect species will attack both hardwoods and softwoods, but preference of one particular wood species is not uncommon; many insect species can live on all woods. Insects appreciate variety in their diet and supplement the wood-food with the natural glues in plywood, attacking such plywood faster than lumber of the same species. Plywood and other wood-base panel products bonded with synthetic resins resist insect attack at least as well as solid wood of the same kind, because these resins are indigestible.

Damage

The largest wood-degrading insects measure several inches in length and excavate holes up to 1 in. in diameter, while the smallest species bore barely visible *pinholes*. Each species arranges the holes in a characteristic fashion—in various regular patterns, or at random. Some species leave round, others oval holes, but for laymen and lumber grading rules they are all *wormholes.* Many insects push the bore dust and/or their excrement to the wood surface; others pack the stuff densely inside the holes. Degree of damage depends on the number and size of the holes or on the extent to which they impair strength, permeability, and appearance. Some insects riddle the whole piece, converting it to powder or leaving only a thin shell intact.

One of the better known destroyers in North America is the carpenter ant, a group living as social insects in communities and comprising several species. Carpenter ants are black or brown, wingless except when mating, and almost ½ in. long; they use wood only for shelter for their nest. Attacking porch columns, window sills, studs in bathroom walls and the like, the ants may destroy such items within a few years.

Identification and Eradication

Whenever wormholes of insects cause alarm, first identify the offender. Countermeasures are most effective if tailored to the particular insect. The type of damage betrays the species, but for laymen there are too many types, and far too many for this book. Here only termites can be considered in detail. For other insects the interested reader is referred to the literature (Nicholas, 1973, Vol. 1). In many cases of infestation the only visible signs of damage are flight holes through which the insects have already escaped, an indication that no further destruction is imminent.

Insects are hard to fight inside wood because insecticide powder cannot penetrate. Kiln drying and fumigation kill them, of course. To prevent the infestation from the start, use either resistant heartwood or treated material. Finished products are safe from most wood-degrading insects because the finish confuses the egg-laying female in search of plentiful sources of food for her brood-hatchery. However, the finish does not bother insects which have penetrated earlier and are growing inside, until finally they bore escape holes.

10.7. TERMITES

Nearly 2000 termite species have been found in the world; they destroy within the United States, tropics, and subtropics more construction wood than all other insects together, bringing whole buildings down within a single decade. That termites can do so much damage is because they obtain nutrition from the wood itself. Most species digest swallowed wood fragments by means of gut microorganisms which secrete an enzyme that breaks down cellulose into water-soluble components, ready to be absorbed for oxidation in the termite body.

Another reason for the efficiency of termites is that they live as social insects in highly organized colonies, in which thousands and sometimes more than a million individuals work together altruistically for the colony's benefit, specializing in certain functions, serving the colony just as individual cells of the human body serve the whole person. A colony may be considered one solid organism, even though it consists of numerous individuals.

Termite Life and Castes

In the colony, *queen* and *king* live in the royal chamber producing fertilized eggs for the brood chambers. The larvae hatching from the eggs develop into individuals of three *castes*. A few become queens and kings of the *reproductive* caste which—if not needed on the throne—fly away, mate, lay eggs, and start a new colony. By far the majority of larvae grow into *workers* who repair and enlarge the nest, search for food, and care for the brood as well as for the two other castes. Some larvae turn into *soldiers* which guard the royal couple and defend the colony against marauding ants. Interestingly, soldiers cannot care for themselves; workers feed them by vomiting into their faces, neither party seeming to mind.

The three castes differ in appearance. Queen and king resemble other adult insects—are dark colored, can see, and have temporary wings for flying out to establish new colonies. Workers and soldiers remain underdeveloped sexless creatures which are blind and pale. The soldiers grow huge heads—not for thinking but for powerful jaws to fight with. Workers have soft bodies and measure only a few millimeter in length.

Termites achieve three-caste differentiation through growth-control hormones, which reproductives and soldiers secrete through their skin to be licked off by the workers (it is not known how that works—fortunately, imagine what man would do with the knowledge). Termites even have the ability to convert grown workers to soldiers in times of need, but so far they have not learned how to make workers out of soldiers.

Their whitish appearance and ant-like social organization has earned termites the name *white ants,* though they lack the thin waists of ants (Fig. 10.6*a*) and are certainly not ants. White color and body softness signify high sensitivity to light and dry air. Termites shun light; they labor in the dark, secluded from open air in their own mostly humid atmosphere. Tunneling into wood they leave a thin surface shell which protects them from desiccation as well as from enemies, and also hides destruction until man probes beneath the wood surface with some pointed tool.

When gnawing wood, workers of some species tear off tiny particles with mandibles (Fig. 10.6*a*), others saw through wood with toothed jaws. Most prefer moist wood for its softness and the humid air within it. Decaying wood suits them best, since it is extremely soft and already partly *digested*.

Subterraneans

Sensitivity against drying and need of moisture determine the living habits of *subterranean* termites, the farthest distributed and most damaging group, which maintains nests in the ground where workers lock up the royal couple and nurture the larvae. Workers venture up to gather wood as food, ascending through posts, framing members, door jambs, and the like; they excavate wooden traffic tunnels to commute between the nest and even upper stories

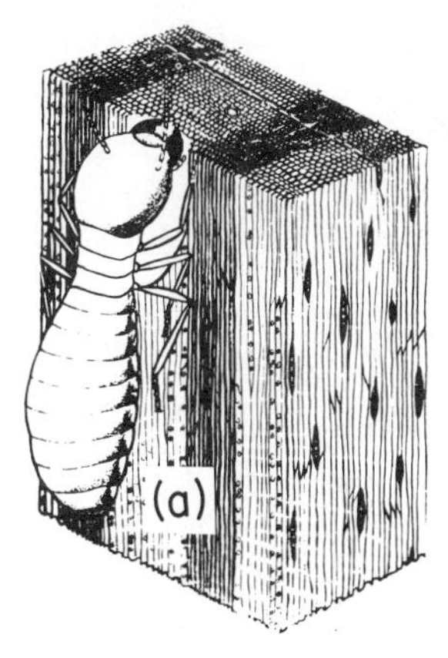

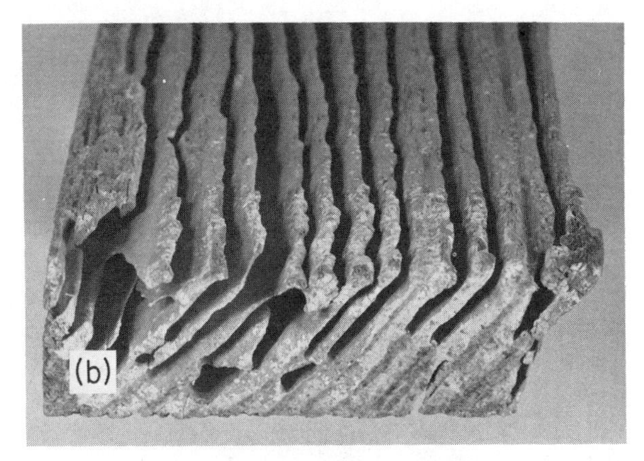

Figure 10.6. Destruction of wood by termites. (*a*) Worker attacking Scotch pine. (Photograph courtesy of G. Schultze-Dewitz.) (*b*) Southern yellow pine honeycombed; the visible surfaces of the piece were inside a wall.

of buildings. Besides the surface shell of wood, the creatures leave thin layers inside wood, typically sparing the dense latewood layer, so that the destroyed piece appears riddled with galleries (Fig. 10.6*b*) and serves the insects as multilane highway.

Subterranean termites first attack lumber through its contact with the ground; they reach wood on foundations above ground either through cavities in the foundation or through their earthern *shelter tubes* (*runways*) made on the foundation surface. Though the nest is not exposed to cold, subterraneans still perfer warm temperatures and do their greatest damage in the South, but they occur in all northern states except Maine, Minnesota, the Dakotas, Wyoming, Montana, and Alaska. These particular states, Canada except Vancouver and Toronto, and Europe from the Alps northward are too cold for them. Subterranean termites do not spread fast because individuals transported in lumber to other places cannot start new colonies without a queen and king.

Subterranean-termite attack can be prevented by denying them access, keeping wood off the ground on impermeable barriers such as a solid concrete foundation, which can be inspected for shelter tubes and cracks. The ground around buildings should be kept free of building debris to deprive the insects of food. After an occurrence of infestation, exterminators treat the ground surrounding the building with insecticides. Workers isolated in the building cannot survive.

Nonsubterraneans

Nonsubterranean wood-inhabiting termites build their nests in wood and live without ground contact, without outside moisture. Having no underground space for depositing waste, they push fecal pellets out of the wood into the open where the pellets accumulate in little heaps, a telltale sign of infestation. This kind multiplies less rapidly than the ground-nesting species and causes less damage, but preventing them is more difficult, because nonsubterraneans indiscriminately infest not only moist but also dry lumber, also because they spread in transported lumber and in colonizing pairs, which fly in through open windows.

The nonsubterraneans appear in two groups of species: *damp-wood* termites (in the United States in tree stumps of Pacific coast forests) and *dry-wood* termites (in the United States in buildings along the Mexican border, the Gulf coast, the Atlantic coast as far north as the North Carolina-Virginia border, and in Florida).

10.8. MARINE BORERS

Marine borers are animals that attack wood in a marine environment—in brackish waters, especially in warmer parts of the world. The borers use wood in the first place for shelter, boring mainly into harbor pilings. Some kinds derive food from the sea, others have the ability to digest the wood borings, and a third category feeds on microorganisms on the wood surface and in wood. Some marine borers belong to the mollusks, others to crustaceae. Mollusks are soft bodied and boneless; they include clams, oysters, and snails. The crustaceae, known as the class of lobsters and shrimp, lack bones, too, but have a hard outer shell.

Though holes *bored* by the wood-damaging mollusks and crustaceae are usually tiny and at most only 1 in. in diameter, they may amount to complete destruction through their great numbers. Some species perforate the superficial wood layer so much that the scouring action of waves removes the surface layer entirely. On wooden boats an unbroken paint film prevents entry by many borers.

10.9. DURABILITY OF WOOD

Durability is the power to endure and to function. In this broad sense the durability of decorative sapwood veneer in a muggy stain-favoring atmosphere, for example, ends with discoloration within a few days. Wood technologists and builders more narrowly define *durability* as physical integrity and strength. This section examines first the factors that influence durability in the narrower sense, before dealing with individual wood species.

Sapwood Durability

Since all woods consist of the same main compounds, and sapwood essentially contains the same extraneous substances (food reserves) in all trees, its durability from tree species to tree species differs little. The food reserves attract wood-damaging organisms and assist the wood destroyers in getting established. Sapwood is not particularly durable, lasting under adverse conditions only a few years; it resists destruction in living trees for reasons explained in Section 10.3.

In spring some food reserves appear as dissolved sugar in the sap of young water-conducting cells, as we know from sugar maple trees, but normally these reserves are stored as starch in living parenchyma cells. Since the percentage of the living cells decreases with increasing distance from the cambial layer, wood-damaging organisms tend to infect the youngest sapwood first.

Living cells consume some food reserves for respiration; that is why, from an alien organisms point of view, the nutrition value of sapwood in felled trees slowly diminishes during storage, until the cells' moisture content drops below fiber saturation and the cells die. For this reason, timber stored green may be slightly more durable than timber converted and dried immediately after felling.

The amount of foodstuff stored in trees peaks in fall at the end of the growing season and reaches a low in spring after the trees have mobilized their reserves for new foliage and growth. Theoretically spring-felled timber should be most durable, but in spring wood-destroying organisms are also most active. The influence of felling weather overshadows the effect of variations in food reserves. Hence winter-felled timber, which dries before summer, is more durable than spring- and summer-felled wood; the difference however is trivial.

Heartwood Durability

As already explained in Section 1.4, many heartwood deposits are toxic to wood-damaging organisms and so prolong wood's lifetime. In the living tree the toxic deposits alter with age and slowly deteriorate, with the effect that heartwood durability decreases somewhat toward the tree's pith. At a given distance from the pith the amounts of toxic deposits frequently diminish from base to trunk-top, where in drier more windy air of fluctuating temperature the tree is less endangered by enemies.

In species with durable heartwood, durability of the trunk as a whole depends on the *heartwood-to-sapwood* ratio, which in turn varies with growth rate and trunk diameter. At the same diameter, fast-grown trees have relatively wide sapwood. Trees in *old-growth* stands generally grow slowly under the canopy of taller trees, forming a trunk which consists mainly of heartwood and is more durable than the trunks of typically fast-growing *second growth* in previously harvested forests. The heartwood of pieces which also contain sapwood is not as durable as pure heartwood pieces, since sapwood serves as a bridgehead for damaging organisms.

Density, Size, and Resin

Wood-destroying organisms penetrate low-density wood relatively quickly for three reasons: (1) thin cell walls offer little resistance, (2) large cell cavities provide sufficient space for diffusion of required oxygen from the outside, and (3) porous wood readily absorbs the needed water. Durability improves with increasing density among wood species and within each species, but not as much as laymen think. Aspen for example deteriorates not much faster than the almost two times denser beech. Density is a poor guide to durability because toxic heartwood deposits far overshadow its influence. The denser pieces of a species are truly more durable only if the higher density results from deposits of toxic substances.

How long wood resists attacking organisms depends on its size, since the attackers need time to penetrate. In some cases, oxygen deficiency in deep layers retards the penetration. In other cases the organisms invade the piece slowly at first, but destroy it quickly after establishing themselves, regardless of thickness. Hence the lifetime of wood increases sometimes less, sometimes more than in proportion to its thickness.

Resin in living trees flows to wounds and seals the openings. After felling and especially during drying the resin becomes viscous and gradually loses its ability to flow, but still obstructs many cell cavities. Resin serves as a physical barrier against penetrating organisms, also retarding water penetration, working in these two ways against attackers. Thus resin protects to some extent even though it is not appreciably toxic.

Durability of Individual Species

Since wood is destroyed by many different organisms acting at different rates, we should define at least three kinds of durability: one with regard to decay, one for insects, and one for marine borers. The resistance to decay has most importance; generally fungi destroy wood faster and in greater quantity than do insects, whose action is rather accidental, depending not only on conditions but also on chance. Fungal spores are always present, while wood-destroying insects may or may not be around.

How long wood resists decay depends on climate; it may hold out seven years in Montana but less than two in Florida. For this reason, instead of naming average durabilities in terms of years, researchers ranked decay resistance itself (Table 10.1). This ranking cannot be accurate, due to variations even within a single tree, to the different influence of climatic conditions on various wood destroyers, and to a particular species' resisting one fungus more than another.

The following data give a crude estimate of durability in terms of years. Two-by-four stakes of sapwood and *nonresistant* heartwood, partially buried in the ground like fence posts, fail in Madison, Wisconsin, after about 5 years, while heartwood stakes of *resistant* species last roughly 20 years. Durability

Table 10.1. Heartwood Decay Resistance of Domestic Woods

Resistant or Very Resistant	Moderately Resistant	Slightly or Nonresistant
Baldcypress (old growth)	Baldcypress (second growth)	Alder
Catalpa	Douglas-fir	Ashes
Cedars	Honeylocust[b]	Aspens
Cherry, black	Larch, western	Basswood
Chestnut	Oak, swamp chestnut	Beech
Cypress, Arizona	Pine, eastern white	Birches
Junipers	Pine, longleaf	Buckeye[b]
Locust, black[c]	Pine, slash	Butternut
Mesquite	Tamarack	Cottonwood
Mulberry, red[c]		Elms
Oak, bur		Hackberry
Oak, chestnut		Hemlocks
Oak, Gambel		Hickories
Oak, Oregon white		Magnolia
Oak, post		Maples
Oak, white		Oak, (red and black species)[b]
Osage-orange[c]		Pines (most other species)[b]
Redwood		Poplar
Sassafras		Spruces
Walnut, black		Sweetgum[b]
Yew, Pacific[c]		Sycamore
		Willows
		Yellow poplar

[a]From Scheffer and Verrall (1973).
[b]These species or certain species within the group have higher decay resistance than most woods in this grouping.
[c]Exceptionally high decay resistance.

in Madison roughly equals the average in North America: winters are cold, forcing fungi into inactivity from November to April; in summer spells of pleasant weather alternate with periods of mugginess and rain, so that loamy top soil remains moist.

It is marine borers and not fungi that limit the lifetime of wood in sea water, where wood may last a shorter time than on land. In United States ports, harbor pilings of nonresistant woods may hold up only a couple of years, and even the most durable species are *eaten up* within two decades.

10.10. WOOD PRESERVATIVES

Wood preservatives are chemicals which, when incorporated in wood, protect the material from attack by organisms. Most preservatives work by poisoning the wood, the organisms' food; some types kill on contact, or are absorbed as a gas of slowly evaporating liquids and sublimating solids. Fumigation exterminates organisms that are present without preventing renewed attack; hence these fumes are not considered preservatives. Some incorporated chemicals protect by holding wood moisture contents on low levels, but to qualify as a preservative the substance must have the effect of a poison.

To avoid undesirable side effects of toxic substances, researchers attempt to preserve wood by fixing nontoxic compounds into the wood structure so that fungal enzymes cannot break the wood constituent. These compounds need to modify the wood only slightly, since the enzymes are highly specific, fitting the functional groups of wood constituents in a lock-and-key relationship. But so far preservatives of this type are not in use.

Common wood preservatives fall into two main classes: oils and waterborne salts. The two differ in many respects, most notably in location; oils remain in cell cavities, whereas salts migrate into the cell wall substance where they can be most effective. When considering a preservative, it is quite instructive to know which kind it is.

Coal-Tar Creosote

Most preservative oils originate from coal tar, a black liquid obtained as a byproduct in the destructive distillation of coal for gas and coke. Coal tar itself does preserve wood, but is too thick for sufficient penetration and has to be separated by fractional distillation. Coal-tar creosote is the group of distillates obtained which contains many of tar's toxic compounds. It evolved as the most common wood preservative, dominating in industrial preservation of bulky mass products, particularly utility poles, railroad ties, and marine piling.

Creosote gives the wood a black or brownish color until it weathers out through the years at the wood surface. About one half of creosote's components are oils, many of which slowly evaporate, causing an offensive odor and harming growing plants. Creosote burns the skin on contact. It also (while still unweathered) interferes with paint and other finishes. It furthermore supports fire, forming a very repugnant smoke. Creosote is for these reasons not suitable for buildings. Outdoors these drawbacks make little difference, indeed there is one decided advantage to coal-tar creosote: being an oil it repells water and resists leaching. Wood properly treated with this preservative lasts under full exposure to the elements for 30 to 60 years.

Creosote is a loosely defined product whose composition varies. Wood preservers use different kinds for different items; for some purposes they add coal tar to make creosote thicker, for others light petroleum oil to make it thinner. Viscosity is critical with regard to penetration and distribution, especially because after treatment the liquid migrates slowly downward by the force of gravity. In utility poles this migration leads to preservative accumulating above the pole's end in the ground, fortifying the zone near ground level where untreated wood decays first and bending stresses are highest.

Pentachlorophenol

Pentachlorophenol, a colorless solid substance, is basically toxic to mammals but poses little danger because it does not dissolve in water and only low concentrations are needed to protect wood. To make a wood preservative, manufacturers dissolve flakes or pellets in mineral spirits and market the solution under the name *penta.* Typically the solution consists of 5% pentachlorophenol in heavy petroleum oil. (Do not confuse pentachlorophenol with the related water-soluble preservative, sodium pentachlorophenate).

The oil has little or no preservative value and remains in wood for a long time, interfering with painting and annoying people in other ways. Pentachlorophenol dissolved in the much more volatile liquified petroleum gas (LPG) is free of these shortcomings, since the gas rapidly evaporates after application, leaving wood clean to the touch and the treatment invisible. Penta dissolved in LPG penetrates extremely well; it may even enter cell walls. For some uses the industry dissolves pentachlorophenol in creosote to obtain the benefits of both preservatives.

Penta is available at lumber dealers in convenient small quantities. People use it in yards, though unpleasant vapors of the solvent and toxicity for growing plants impose limits. For some purposes, penta works adequately even when only brushed on, but most products require more expensive treatment. Pressure-applied penta protects wood nearly as long as coal-tar creosote, though not against marine borers.

Water-Repellent Preservative (WRP)

This type is frequently brushed on before painting, as discussed in Section 14.5. Typically, WRP contains pentachlorophenol as its active ingredient and differs from common penta mainly in the small amount of wax added to repell water. Less poisonous ingredients than pentachlorophenol have to be used for wood in contact with food. To permit painting soon after WRP application, manufacturers employ fast-evaporating mineral spirits as solvent; for this reason WRP belongs along with penta to the class of preservative oils.

Millwork factories treat windows, entrance doors, containers, and decking of trucks with WRP. Preservatives of this type can be applied by brushing and by dipping. Pressure treating is less common, because evaporation of the larger amounts of solvent takes too long and causes difficulties in painting subsequently.

Waterborne Preservatives

Numerous salts are toxic for wood-damaging organisms and have found use as wood preservatives in aqueous solution. When dried or redried, as required for most uses, water evaporates while the salt remains in the wood.

Since the salts enter wood in solution, they can be assumed to leach out under exposure outside; being hygroscopic they attract moisture, which as *sweat* water may dissolve them. However, many toxic salts react chemically with wood and lose solubility within a few weeks; some become insoluble once a volatile solubilizing reagent evaporates. Therefore the industry prefers the name *waterborne* to *water soluble.* Nearly all preservative salts on the market do become insoluble and protect the wood as long or even longer than coal-tar creosote, depending on the attacking organisms. Some are applied to fortify creosote for special products such as marine piling.

The chemical industry offers waterborne preservatives as dry mixes, as pastes, and as concentrated solutions. Most brands contain compounds of arsenic, chromium, copper, zinc, and fluoride; many are marketed under trade names such as *Celcure,* which essentially is acid copper chromate. In the literature these chemical names appear in abbreviations, like *CCA* for *chromated copper arsenate.* Typically, waterborne preservatives give wood a green tinge and are therefore called *green salts*; a few appear colorless at first and weather to a driftwood tone. Waterborne preservatives have no objectional odor, and the treated wood is clean to the touch and can be painted or even clear-finished. Its fire characteristics compare favorably with those of untreated wood.

The salts lend themselves safely to dwellings and have made the *All Weather Wood Foundations* from lumber studs and plywood sheathing possible. On basis of weight the prices of waterborne preservatives appear high, but the treatment is relatively inexpensive because such small amounts as 0.5 lb/ft^3 provide sufficient protection. In moist wood the salts once used to corrode metals, until the industry found ways to minimize the damage. It is important to dry the treated wood before installation, to avoid shrinkage as well as corrosion. Another disadvantage of salts is insignificant in many cases: they impair wood toughness.

10.11. PENETRATION OF PRESERVATIVES

The effectiveness of wood preservatives depends on their penetration and retention. Penetration is expressed in units of distance from the wood surface, or in terms like *all the sapwood. Retention* means *the mass of preservative in a unit volume of wood* (for example 10 lb/ft^3), measured immediately after the preservation process or after certain periods of exposure. Large penetration is not necessarily associated with large retention because wood preservers control the two independently through the concentration of dissolved preservatives and through the particular pressure schedule.

Preservatives move through wood mostly as a whole fluid by convection, flowing like water in a hose and driven by pressure differences or by capillary forces. Some preservatives migrate by diffusion, by irregular motion of dissolved preservative molecules through the moisture in wood or through the cell wall substance. It is the concentration gradient which drives salt molecules to penetrate when green wood is brushed or sprayed with salt solution or covered with salt paste.

Penetration naturally depends on wood permeability, a property which has been discussed in Section 1.5. Passage of thick oil requires larger openings than do the dissolved molecules in waterborne preservatives. The latter diffuse even into cell walls, by-passing closed pits and finally reaching the very center of any piece of wood, although this process may take months.

Due to the orientation of fibers and vessels, end grain surfaces and cross-grained lateral surfaces absorb much larger amounts of preservative than do straight-grained lateral surfaces. The fairly rapid radial penetration along the rays causes starshaped patterns of impregnation in cross sections of trunks.

Moisture has a different affect on the penetration of oils versus that of waterborne preservatives, its influence depending also on whether the solution flows or the molecules migrate by diffusion. Free wood moisture blocks the flow of oil more than the flow of aqueous solutions, simply because water and oils do not mix. For diffusion the wood must be wet, since water serves as a pathway for the salt. Bound moisture in cell walls will influence flow through the capillary contact angle, that is, the angle which the incoming liquid forms with the cell walls. Bound moisture holds oils back while easing the way for aqueous solutions; it affects penetration of preservatives also by widening pit openings, so that fiber saturated wood can absorb waterborne preservatives (and possibly also some oils) faster than dry wood.

10.12. PRESERVATION PROCESSES

Penetration and retention are most of all influenced by the method of preservative application, that is, whether liquid is pressed into the wood or simply brought in contact with the surface for various periods of time. Penetration increases along with the costs of application, from brushing and spraying to dipping and soaking, thermal processes, and pressure treatment. Due to various other factors that influence penetration, only approximate penetrations can be given for each method.

Brushing and Spraying

In these two methods of application, preservative coats the surface and soaks in by capillary forces. *Run-off* limits the amount of liquid that will stay on the surface, thus limiting retention and penetration. Wood surfaces can hold relatively large amounts of viscous preservative, but not enough to compensate for their slower flow. Rough-sawn horizontal surfaces retain about two times more liquid than planed vertical surfaces, but the difference between these two surfaces is smaller for penetration than for retention, because loose surface fibers seize hold of liquids permanently and so contribute more toward retention.

A second brushing or spraying will naturally increase penetration, provided the first application has not sealed the surface. Some sealing may result from the fixing and precipitation of salt preservatives, but the fixing process is too slow to interfere significantly if the first brushing is still wet.

Brushed and sprayed preservatives penetrate lateral surfaces a few millimeters at most, protecting wood's very surface layer, enough to retard weathering and preserve very thin wood, but not enough to prevent decay of thick wood for a long time. Brushing and spraying prolong the lifetime of fence posts, for example, no more than one or two years, at the very most five years.

Dipping, Soaking, and Steeping

Dipping the wood momentarily into preservative essentially equals brushing; *smart* molecules at the wood surface notice little difference between both methods, except in the surface checks which dipping is likelier to fill. Some sawmills dip all lumber, passing it through a bath of fortified penta to prevent blue stain during subsequent storage, shipment, and drying. The millwork industry applies this method to window sash, frames, and other products, dipping them about 3 min, which is relatively long but still permits hardly more than 1 mm of transverse penetration. At end-grain surfaces of permeable woods the preservative can soak 2 in. or even 3 in. deep in this time.

Soaking and steeping last much longer than dipping and give higher retentions. The term *soaking* is preferred for immersion less than one day in any kind of preservative, whereas *steeping* refers to longer immersions in salt solutions. Steeping leads to complete penetration if it lasts long enough, as we may conclude from how wood in lakes ultimately sinks to the lake bottom. Soaking and steeping lend themselves to home treatment of fence posts, by setting the posts upright into a drum filled with the liquid, leaving unsoaked the upper part which does not need protection anyway. Posts of pine, a genus with relatively permeable sapwood, if steeped just 2 days will last up to 20 years.

In the *double-diffusion* process green wood is steeped first for a few days in one solution and then a few days in a second. The two chemicals react with each other to form an insoluble preservative. Compared with steeping in a single waterborne preservative, double-diffusion should go deeper, since the one-dip process tends to close pits and capillaries in cell walls when reacting with wood, whereas in the latter process the wood is supposed to remain open. Sodium fluoride and copper sulfate are the most practical chemicals for this treatment; both cost little and are kept in stock by chemical dealers. Because copper sulfate is corrosive to iron even if painted, it should be applied in plastic containers such as garbage cans or wooden barrels.

Thermal Processes

Heat is useful in various ways to force the preservative deep into wood. The *hot-cold soaking* method consists of submerging the wood first in a hot bath, then transferring it quickly to a cold one. The cooling causes air in cell cavities to contract, creating a vacuum that pulls the preservative in. The heating and cooling periods are determined by the penetration of the heat and cold wave, respectively, which in the case of fence posts takes a few hours (Section 8.1). Retention increases with higher hot-bath temperatures, because hotter air contracts more when cooled and because heated preservative is less viscous. Also, in hot wood some moisture evaporates, pushing air out of the cell cavities, and upon condensing creates additional vacuum. At peak temperatures near the boiling point of water, the vacuum caused by vapor becomes stronger than that created by contracting air. Hot-cold soaking results in significantly deeper penetration than cold soaking, and has its application where cold soaking is inadequate.

When wood is heated in oils or concentrated aqueous solutions of salt to above the boiling point of wood moisture, steam pushes all air out of cell cavities; during later cooling within preservative the condensing steam creates a nearly full vacuum and pulls preservative into all cavity space. Free wood moisture boils near 212°F, while bound moisture has elevated boiling points which increase with decreasing moisture contents, as detailed in Section 8.2. Moisture boils first at the surface until the surface temperature rises and moves the drying front inward. Steam replaces air and free water between the surface and the drying front, so that during subsequent cooling the preservative fills the cells at least up to that front. Complete penetration and maximum retention require boiling out all free moisture. Such *high temperature drying in preservatives* was practiced already in the last century, but is still hardly known; it has its place where wood should be dried and treated with inexpensive equipment.

Pressure Processes

Dipping, soaking, and the thermal processes involve pressures up to one atmosphere, while real pressure processes work at around ten atmospheres. Wood preservation plants pressure-treat in huge horizontal cylinders, sufficiently long for utility poles and wide enough for loads (*charges*) carried on trams.

The treatment cylinders withstand the pressure and vacuum of any desired combination of penetration and retention. In the *full-cell* process, air is drawn out of wood before creosote is pressed into the emptied cells to achieve highest possible retention, needed for marine piling. In the *empty-cell* process, for low retention, air is left in the wood; upon release of pressure the expanding air cushion pushes most of the preservative out but leaves walls of the cell cavities coated. In both processes a final vacuum frees the charge from dripping and reduces *bleeding* or *blooming*, which occurs especially when solar radiation heats the wood during its service. Special treatment schedules, involving a final steaming, can eliminate bleeding altogether when a *clean* product is needed. A consequential reduced retention in the surface layer, which is the front where the fungus attacks, can be tolerated because preservatives tend to migrate surfaceward anyway, where volatiles evaporate and the active ingredients are deposited.

Since even high pressure does not open aspirated pits, preservatives penetrate the heartwood of coast-type Douglas fir and of southern yellow pine (the most important species) only 1/4 to 1/2 in. The durability of poles depends then on retention in the sapwood ring and on natural heartwood durability. In the spruces (except Sitka spruce) and some other species, pits also aspirate in dry sapwood. To achieve sufficient penetration regardless, the industry incises the surface of poles in this category in such a way that preservative reaches nearly all the sapwood in fiber direction. Pit aspiration affects longitudinal permeability as well, but the preservative penetrates even aspirated wood much farther lengthwise than transversely.

Europeans treat large poles of spruce in the green state before sapwood pits can aspirate. They lay poles on the ground and press waterborne preservative into the butt end until the solution appears at the top end. Preservative flows in a pipe down from a tower-supported tank into a cap which is tightly attached to the butt end, pushing sap out of the sapwood cells and replacing it with preservative. The method is known as the *Boucherie* process.

Practically all utility poles, railroad ties, and marine pilings are pressure-treated and with proper penetration and retention they last for 30 to 60 years. Treated fence posts, lumber for porches, ties for retaining walls, and other treated products can be procured from lumber dealers. Not all of these products are pressure-treated and not all need to be, but buyers want pressure-treated material if they have paid for it. Dealers buy from and rely on the wood preservation industry, which in turn maintains an inspection service, the American Wood Preserver's Bureau (AWPB) of the American Wood Preservers Association (AWPA). The Bureau grants qualifying member-plants the right to stamp an AWPB quality mark on their products. The mark guarantees that the right kind of preservative has been used and that penetration as well as retention meet particular needs.

The AWPB requirements are especially rigorous for lumber and plywood of *All Weather Wood Foundations*: the preservative retention must be 50% higher than for other woods in ground contact, fewer preservatives are permitted, and each charge must be approved before shipment. With foundations of such preserved lumber and plywood, properly constructed homes outlast the home owner in any case.

11

Working Properties of Wood

The many findings of researchers and craftsmen about woodworking have been compiled in the *Wood Machining Processes* book by P. Koch, in engineering handbooks, in thousands of journal articles, in texts for wood crafts, and in woodworking machine manuals. This chapter shows how wood properties influence woodworking, without going into detail more than necessary on machines and tools.

Wood is a relatively soft material that offers little resistance to indentation. To some extent tools press it aside instead of cutting it off. The compressed elastic material also pushes back against the tool, creating friction and significant resistance to feeding the piece into the machine tool. Later the compressed tissue recovers and causes surface roughness.

High tool velocity minimizes compression, as this simple experiment illustrates: hold a sheet of paper in the air and try to cut through it with a razor blade in slow motion—the blade does not cut, it rather pushes the sheet away; but at high velocity the blade does strike through since it leaves the sheet no time for evasion. The material's inertia gives the effect of greater rigidity. Wood softness calls for fast machine tools to bring about smooth cuts, reduced friction, minimized power consumption, and easy movement of the workpiece into the tool, for saws as well as planers.

11.1 SAW BLADES IN RELATION TO WOOD

This section applies to all types of saws—circular saws, band saws, and reciprocating saws—all used for similar purposes. Within each type, the saws differ in the width of the kerf they cut and in tooth shape, as needed for different tasks, various kinds of wood, and the wide range of wood moisture contents.

Side Clearance

Saw teeth are supposed to cut a kerf for the saw blade to move freely. However since to some extent the teeth push wood aside and compress it, the blade may rub on wood. A theoretical kerf cut by teeth exactly as thick as the blade would be too thin. Heat of friction raises the blade temperature and some zones of the blade expand through heat more than others, causing the blade to warp. Warping in turn results in additional friction, intensified heating, blade oscillation, and charring of wood, forcing the machine finally to a halt.

Actually saw friction has additional causes. Dull teeth cut incompletely and tear up the soft surface. Some fibers and fiber bundles remain connected with the solid tissue, making the surface fuzzy. This fuzziness as well as sawdust between blade and wood have the same effect as wood compression and call for wider kerfs.

The blade's *cutting edge width* (Fig. 11.1) must exceed blade thickness to give side clearance and is provided in most blades by tooth *setting.* The teeth are either alternately deflected (*spring-set,* Fig. 11.*1a*) or spread at the tips to both sides (*swage-set.* Fig. 11.*1b*). The required set depends on the degree of wood compression, with highly compressible light species and green lumber requiring larger sets than heavy dry wood. Since ice in cell cavities resists compression, water-soaked frozen wood needs the least set. Saws have their optimal side clearance when consuming the least power. Excessive clearance costs extra power to convert more wood than necessary to sawdust, while inadequate clearance raises power consumption through friction.

The actual clearance varies with the saw blade size, which in turn is determined by the depth of the cut. Consider first the huge circular saw

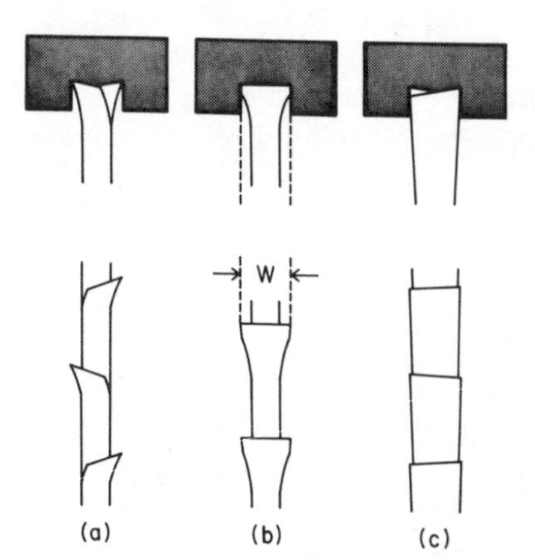

Figure 11.1. Three types of side clearance for saws. (*a*) Spring-setting, (*b*) swage-setting, (*c*) hollow-ground blade. *w*, Cutting edge width.

used in sawmill headrigs for converting sawlogs into lumber. Blades of this kind reach the log's opposite surface and should have a diameter roughly twice that of the log. In this size they deflect easily and deviate from the desired path; they must therefore be adequately stiff—some are $^{1}/_{4}$ in. thick. Since the wide cutting edge width involves large compression of wood and even a slight deviation of the huge blade causes excessive friction, the teeth need large sets: for softwood up to $^{5}/_{64}$ in. on each side, for hardwoods about $^{4}/_{64}$ in., and for frozen wood $^{3}/_{64}$ in.

European sash gang saws, designed for small sawlogs and gaining rigidity by being stretched in a sash frame, are only $^{1}/_{16}$ in. thick and have small sets, $^{3}/_{128}$ in. for green softwood and $^{2}/_{128}$ in. for frozen logs of any kind. Small circular saws in home workshops cut green softwood satisfactorily with $^{2}/_{128}$ in. and dry hardwood with $^{1}/_{128}$ in. set at each side.

Set teeth work with pointed tips rather than with sharp edges, producing uneven grooved surfaces. Hollow-ground circular saws gain side clearance by means of a thick rim and teeth that are thicker throughout than the main body of the blade, as indicated in Fig. 11.*1c.* Lacking distinct protruding points, these *planer* saws produce relatively smooth surfaces.

Types of Saw Teeth

The tooth type determines energy consumption, how smoothly the saw cuts, how safely it works, and the feed-in rate. Blades for fine cuts on plywood, for example, have many small teeth designed to take tiny bites without tearing out splinters from poorly bonded veneer. The bite, or the amount of wood cut off by each tooth in one pass, is crucial with regard to smoothness and power consumption. For lengthwise cutting (*ripping*) of thick lumber with relatively little power, saws have large teeth which take big bites to be carried out of the kerf in spacious gullets between the teeth. The gullet space should be at least 2.5 times larger than the bite's solid wood volume.

Figure 11.2 shows major tooth forms. Hook-shaped types (D) are most efficient. As the rake angle increases, the cutting resistance significantly decreases, up to 20°, at which angle the decrease levels off. The rake angle also affects the force required to feed a workpiece into the machine. In circular saws, which typically turn against the approaching workpiece, the infeed power drops as the angle increases, so that around 30° the saw almost feeds itself. From about 40° up the blade deviates and the teeth *dig into the wood,* producing rough surfaces and wearing fast, all reasons to limit the angle to 30 or 40°.

For many tasks, tooth strength and tooth stiffness impose even lower limits. The hook-shaped tooth type D is more likely to break out and to deflect than the sturdy types A, B, and C. Therefore most saws for hard dry wood have rake angles of only 10°. Spring-set teeth work with relatively small rake angles because they must resist one-sided stress and deflection.

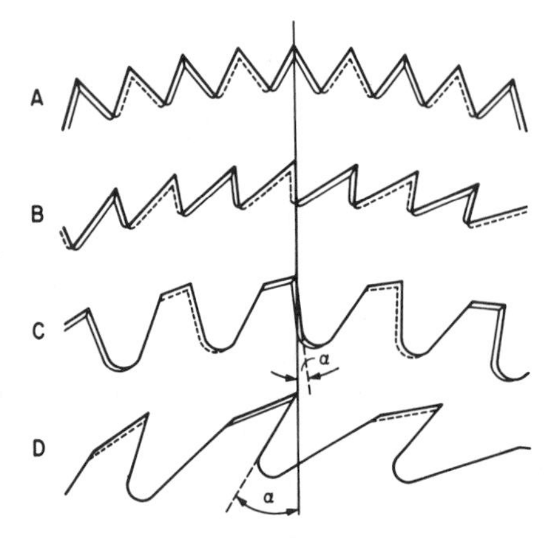

Figure 11.2 Types of saw teeth. α, The rake angle, is negative in type C. (By C. H. Hayward.)

Like a knife thrust into wood in fiber direction, ripsaws meet relatively little resistance and can afford to have weak teeth with larger rake angles than crosscut saws. In Fig. 11.2, type D is for ripsaws which make long cuts and should work efficiently. The crosscut saw types A, B, and C sacrifice efficiency in favor of strength, stiffness, and wear resistance. The more universal crosscut blades can be used for ripping, but require excessive cutting power and feeding force, whereas crosscutting dense wood with ripsaws may damage the teeth and give very rough surfaces.

Deviation and Pinching

Saws have two rows of tooth tips, one on each side of the rim. When cutting along a hard knot, the row at the side of the knot meets more cutting resistance and is subjected to more side pressure than the other row, with the effect that knots push the blade out of its course. Since dull teeth cut less well and are subjected to more pressure by the compressed wood than sharp teeth, dullness contributes to deviation and may appear as its cause. For the same reason, excessive feeding force and inadequate side clearance push the blade out of line. One-sided pressure against the blade's side may be caused also by variations in fiber direction, but the resulting difference in pressure is rarely large enough to force the blade to one side whereas one-sided tooth setting and one-sided sharpening lead to deviation inevitably.

Some pieces of lumber are internally stressed, their surface being under longitudinal compression and their core under tension, or vice versa as explained in the Drying Section (3.9) on casehardened lumber. Ripping upsets the equilibrium between tensile and compressive forces, so that the pieces bend and pinch the saw blade. Most internal stresses in lumber result from moisture gradients and from casehardening; others, causing difficulties mainly in sawmills, develop in the growing trees, either as symmetric *growth stresses* or as asymmetric reaction wood stresses (Section 2.5).

11.2. CHIPPED, RAISED, AND LOOSENED GRAIN

Chipped and raised grain are common planing defects that occur also during molding and other kinds of dressing, all of which essentially amount to planing operations using short curved knives. Loosened grain frequently evolves from raised grain and develops on wood during use. To control these defects it is essential to understand how they come about.

Chipped Grain

Planer knives that work lengthwise may split rather than cut, as illustrated in Fig. 11.3, in which the split typically spreads ahead of the knife in direction of the fibers. Splitting does not matter where the fibers *run* either parallel to or *out of* the surface, but where they *run into* the workpiece, as illustrated, the split spreads below the desired planing line, until the tool finally breaks the splinter out. With further advancement of the workpiece, the tool again engages with solid wood, splitting and breaking off more splinters, so that short narrow spots alternate in a staggered fashion as shown. Chipped grain (*grain* in the sense of fiber) is defined as a surface chipped or broken out in very short fragments below the line of cut.

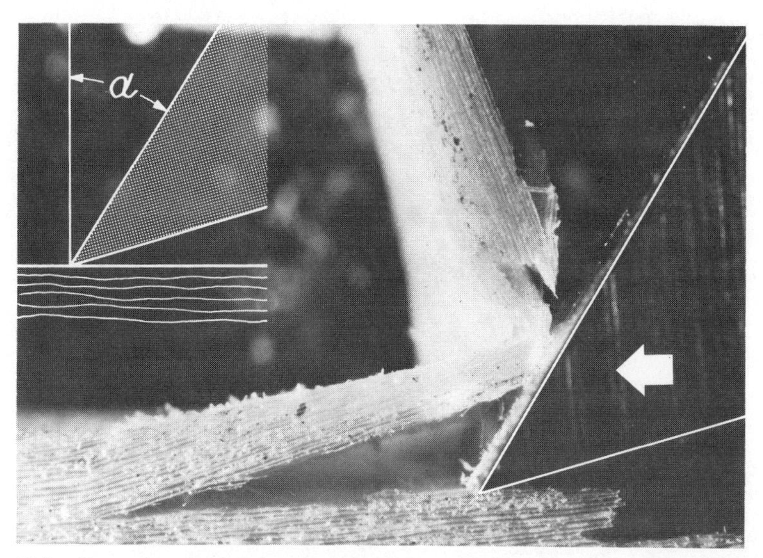

Figure 11.3. Formation of chipped grain when cutting against the grain. α, Rake angle. (From N. C. Franz, *An Analysis of the Wood-Cutting Process*, Ph.D. thesis, Univ. of Mich., Ann Arbor, 1956.)

Theoretically, chipped grain can be avoided by planing cross-grained wood always *with* the grain. However, this is impractical because in many cross-grained pieces the fiber direction alternates. Most cross grain results from knots and varies around each knot, so that the fibers may *run in* ahead of the knot and *run out* behind it or vice versa.

Planer knives naturally split less when forming a small rake angle (Fig. 11.3). If the wedge-shaped knife in the figure would stand upright perpendicular to fiber direction, forming a negative rake angle, it would scrape and cause no splitting at all, while a knife pointing in fiber direction would split the most. The rake angle controls chipped grain.

Figure 11.4 illustrates in the example of a wooden hand plane additional ways for controlling chipped grain. The pressure lip left of the shaving and the narrow mouth hold potential splinters down, in this way restricting propagation of splits. The chipbreaker has the same effect as it bends the shaving forward. Pressure lips and chipbreakers appear in principle also in the revolving tools of machine planers, though frequently under different names.

Raised Grain

Raised grain (*grain* in the sense of annual rings) is an uneven surface in which hard latewood is raised above earlywood but not torn loose from it (Fig. 11.5). Planing raises the grain at the surface of flat-grained lumber in species whose latewood is much denser than earlywood, especially among softwoods (Fig. 11.6). At the pith side surface, planer knives pound the dense latewood tongue into the soft bed of porous earlywood, indirectly crushing the earlywood. Later the earlywood recovers like a compressed cushion and pushes the latewood tongue above the surface. The planer knife does not cut off the latewood tongue, for the same reason that a footstep does not break a nut in sand.

Raised grain rarely occurs at the bark side surface, where the tip of the latewood tongue consists of relatively soft latewood supported by relatively hard earlywood (Fig. 11.6*c*). On the bark side surface, at the point of sharp earlywood-latewood contrast, dense latewood supports porous earlywood of the more recent annual ring, which the knife easily cuts off, like a footstep

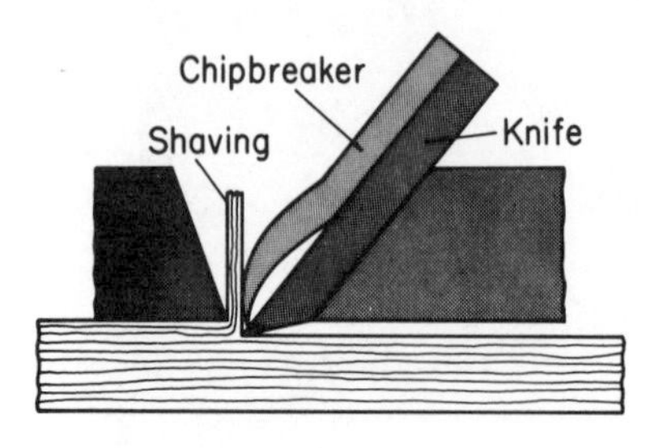

Figure 11.4. Hand plane in action. (By G. Pahlitzsch.)

Figure 11.5. Pith side surface of white spruce with raised grain. (From Forest Products Laboratory, 1965.)

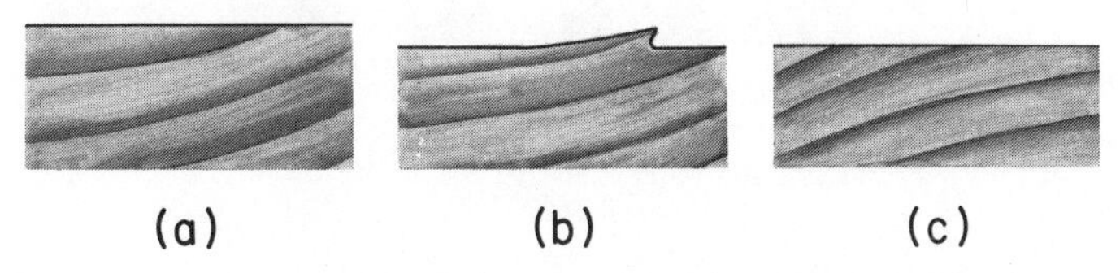

Figure 11.6. Occurrence of raised grain in softwood with gradual transition from earlywood to latewood. (*a*), (*b*) Pith side surface before and after planing, respectively. (*c*) Bark side surface.

crushing a nut on rock. Sudden inward transition from hard to soft promotes raised grain, but not from soft to hard.

Swelling and shrinking after planing accelerate recovery of the crushed earlywood and accentuate the flaw. Sharp knives obviously cause no raised grain or only a little, while planing by hand rarely raises the grain because it involves little pressure.

Loosened Grain

In loosened grain dense latewood tongues rise out of the flat-grained pith side surface as in raised grain, but they stick out farther as the whole sheath of an annual ring separates from the tissue beneath and rises; we say the wood *shells* (Fig. 11.7). Loosened grain frequently starts at the exposed latewood edges of raised grain when wood swells and shrinks. Dense latewood tends to swell and shrink tangentially much more than light earlywood, creating tangential shear stress and finally causing separation at the interface between latewood and crushed earlywood of the annual ring underneath.

Figure 11.7. Loosened grain in, left to right, southern yellow pine flooring, soft maple flooring, southern yellow pine dowel, and white ash dowel. The dowels failed due to excessive turning tool pressure. (From Forest Products Laboratory, 1965.)

Some grain is loosened already in the planer by the pounding of dull knives and by feed rolls. Footsteps crush the underlying earlywood of floors, where loosened grain is most common. Heavily used benches and tables exhibit the defect too. Like raised grain, loosened grain occurs predominantly at the pith side surface where the annual rings *run out* of the tissue and form distinct tongues. The sharp slivers at the loose tip of the projecting shell render cleaning difficult and contribute to accidents.

11.3. WORKING QUALITY

The term *working quality* covers cutting resistance, smoothness of the worked surface, and tool wear—characteristics that have great significance for a material like wood, which does not lend itself to plastic forming and must be shaped by means of cutting.

Cutting Resistance

The resistance or force of cutting determines what strength woodworking machines have to be designed for and how much power they consume. In hand woodworking the cutting resistance is a major factor in choosing the wood species.

Influence of Density and Direction Some wood species are much easier to saw, plane, rout, turn, bore, cut to veneer, sand, chip, grind, and fiberize than others. The ease of working naturally depends in the first place on density; the cutting resistance increases sometimes in proportion to density, sometimes faster than that.

As mentioned before, it takes relatively small force to thrust a knife blade into wood in fiber direction. Wood works lengthwise more easily than transversely. This transverse orientation is one reason why knots resist cutting, the other being the high density of knots. In ripping Douglas fir lumber, for example, knot tissue offers about three times more cutting resistance than clear wood tissue.

The effects of density and fiber orientation give an estimate for the working of plywood, particleboard, and fiberboard. Due to their alternating fiber directions, the cutting resistance of these panel products falls between that of ripping and crosscutting lumber of the same density. Synthetic binders add some resistance, as explained below under *wear of tools.*

Influence of Moisture Content To understand the influence of wood moisture, keep in mind that most saws, planers, and other tools work in different ways. Sharp tools separate the wood at the edge with a force that depends on wood hardness; after some wear they start to split and shear the wood apart, overcoming tensile strength perpendicular to the fiber direction and shear strength, respectively. Dull tools may crush the tissue, as when wood breaks under compression in fiber direction. The average working resistance depends therefore on wood hardness, tensile strength, shear strength, and compressive strength, all of which decrease with increasing wood moisture. Consequently the resistance should become lower as the moisture content rises. For the following four reasons, however, moist wood is not always worked more easily than dry wood.

1. Acceleration of heavy wet chips by fast tools requires more force than acceleration of light dry chips.
2. Moist wood, being tough and sustaining large strain before failing, splits less than brittle dry wood and must be crushed or cut, which requires more power.
3. When working in soft moist wood, tools tend to compress rather than cut, resulting in friction.
4. Being tough and compressible, chipped moist wood hangs together in large shavings which stick to the tool and increase friction.

Influence of Temperature Though heat generally affects properties of wood like moisture does, it always seems to facilitate cutting. It is characteristic that the plywood industry heats logs of dense wood species for veneer cutting. Freezing as a rule sharply increases the cutting resistance of wet wood, due to ice in cell cavities. However, in some sawing tests frozen wood required less power, possibly because ice also embrittles the tissue.

Surface Smoothness

Smoothness is besides accuracy the ultimate quality criterion of woodworking. Many working steps do not achieve the smoothness desired for the final product and make further processing necessary. Sawn surfaces are too rough for most purposes and require planing, while planed surfaces in turn may have to be sanded with ever finer abrasive paper until the piece is smooth enough.

The well-known roughness of end-grain surfaces and chipped grain illustrate that the smoothness of worked surfaces depends on fiber orientation. Cross grain causes rough surfaces; in the form of irregular grain and interlocked grain it impairs the workability of many tropical wood species.

To reach the same smoothness, highly compressible light woods require sharper tools than dense woods. For balsa, the lightest wood species, tools must be extraordinarily sharp, as we may conclude from the crumbling of styrofoam when cutting it (if the process can indeed be called cutting) with a fairly sharp knife. Using the same tools, denser woods generally work smoother.

On sanded surfaces, the scratches made by the grits of the abrasive paper should be invisible. At lateral surfaces, sanded in fiber direction, the size of wood pores determines the size of allowed scratches and the size of sandpaper grits. Fine pores require fine sandpaper at the finish.

Tools press partially detached fibers into the backing tissue, particularly into light moist wood and into large pores of hardwoods. Later the fibers stand up and appear as a kind of *raised grain* quite different from the raised grain described in Section 11.2. When sanding moist lateral wood surfaces in fiber direction, sandpaper grits tear up long fiber bundles and individual fibers without cutting them off. These projecting shreds curl and make the surface fibrous, causing *fuzzy grain.* The defect also occurs in sawing and planing with dull tools which do not cut through uptorn fibers.

Tension wood fibers are too tough for many cutting tools, and they fuzz so much that fuzziness has evolved as an indicator of tension wood (Section 2.5). Among the native woods, cottonwood and aspen fuzz most because they contain large amounts of tension wood in addition to being soft. Softwoods show relatively little fuzzy grain.

Wear of Tools

Tips of saw teeth, edges of planer knives, and the grits of sandpaper are much harder than any wood species, but still become blunt through thousands of engagements. The wear naturally depends on the difference in hardness between tool and wood, hard dense wood naturally dulling the tools fastest. Several tropical wood species are much worse in this regard than would be expected on grounds of density, because their cell cavities contain tiny crystals and other mineral deposits which rapidly wear tool edges down, without

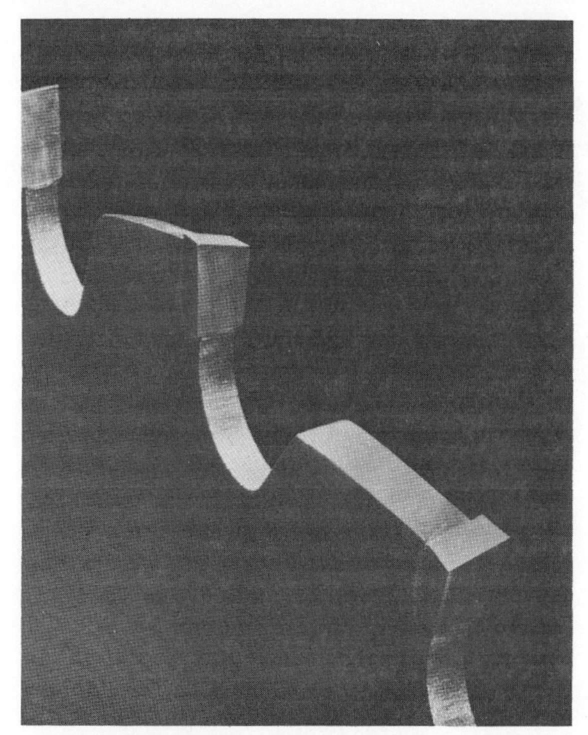

Figure 11.8. Carbide-tipped saw blade. (Photograph courtesy of Simonds—Wallace Murray.)

however affecting wood density and hardness, as measured by pressing a steel ball into the surface.

The urea resin in particleboard and in plywood is harder than wood and accelerates tool wear, especially in particleboard which contains more resin than plywood. Phenolic resin in particleboard for exterior use and in Exterior plywood is softer than urea resin and dulls tools less. Fiberboard compares favorably with particleboard of the same density in this regard, because it either lacks resin or contains only the phenolic type. With the exception of casein, natural glues cause no more wear than wood does.

Lumber and wood-base panels that cause excessive wear are worked most cheaply with carbide tools. Carbide is a compound of metal and carbon, fused together in heat to an extremely hard material that has to be ground with diamond-impregnated wheels. The compound is too brittle, too expensive, and too hard to shape for whole tools; therefore tool manufacturers solder or weld carbide tips to tool bodies made of steel (Fig. 11.8).

12

Solid Bending and Transverse Plasticity of Wood

Curved pieces sawed out of solid lumber are cross-grained and therefore weak, besides being a waste of material. Laminate bending (Section 5.3) gives strong products, but involves gluing under pressure which in curved pieces is difficult to apply, and in some cases laminations distract from the appearance. Solid bending avoids these shortcomings.

Bending consists of lengthwise compression and stretching, as explained in Section 4.1. Permanent transverse deformations have already been mentioned, for wood-base panels and case-hardening, yet these are small compared with transverse compressions, applied in order to densify wood for exceptional strength. Solid bending and densification follow the same concepts, which are discussed in this chapter together with related wood properties.

12.1. PLASTICITY

When attempting to bend a thin stem into the shape of a walking stick, we face two problems: the stem breaks before reaching the desired handle curvature, and on release of force the bend straightens out because it is elastic, not plastic. Lack of plasticity and limited formability render solid wood bending as well as densification difficult. Both plasticity and formability, are closely related as we shall see.

How to Plasticize Wood

In wood only very mild bends are entirely elastic. Sticks that have been bent relatively far do not return completely to the original shape, a minor percentage of bend remains. The plastic percentage increases with an increase in total deformation and with how long the stick is held in shape. To achieve a permanent narrow curvature, bend as far as possible without breaking and clamp the stick for a long time. However on release out of the clamp, still too little of the bend remains. Plastic deformation dramatically increases when bending hot and moist, that is, after boiling wood in water or steaming it. If in addition we arrest the bend until the wood is cold and dry, most of it will stay. The combination of bending far in the hot moist state with cooling and drying in the desired shape produces narrow bends, whose plastic share reaches 80 or even 90% of the total deformation.

We apply the same concept when pressing creases (a sort of 180° bend) into trousers of cloth. The iron provides heat, moisture is either in the fabric already or it steams out of the iron, and we hold the crease until the fabric is cool and dry. For cotton fabrics the similarity to wood is not surprising, since cotton is cellulose, wood's main constituent. In a process invented by this author, the wood industry flattens buckled veneer exactly as fabrics are ironed, that is, by pressing between hot platens at a degree of heat which depends on the moisture content. Dry veneer gains plasticity through relatively high temperatures, while conversely more moisture compensates for lack of heat.

Explanation of Plasticity

In the normal unstressed state, submicrospically small cell wall particles have a certain normal distance from each other, as if held in place by rubbery connecting rods (Fig. 12.1*a*). Moderate tension pulls the particles farther apart, indicated by the elongation of the piece in Fig. 12.1*b*; at this point the rods stretch but hold, the deformation is elastic, and on release of the force the material would return to the state of Fig. 12.1a. High tensile force overstretches some rods, causing particles to slide far apart and a few to slip sidewise into the enlarged gaps between neighbors (Fig. 12.1*c*). Upon release of the force the particles retract to normal distances, though not to the former places; the slippages last, the wood remains elongated or strained, set in tension. Pressure causes a set in compression in basically the same way.

Slippage requires time because individual particles slip when their molecular motion just incidently matches the molecular motion of an adjacent particle; or in other words, when the order of the particles is unbalanced by chance. Heat is molecular motion (Section 3.1) and promotes plasticity for this reason.

The various cell wall constituents swell and shrink differently, just as they expand and contract differently when the temperature changes. Again the difference in movement unbalances the order and promotes slippage, as shaking a sieve drives more sand through the openings. This is why drying and cooling makes bends permanent. The fact that moisture promotes plasticity is explained with the increased distance of the particles, which being separated by water in the swollen wood slip easily.

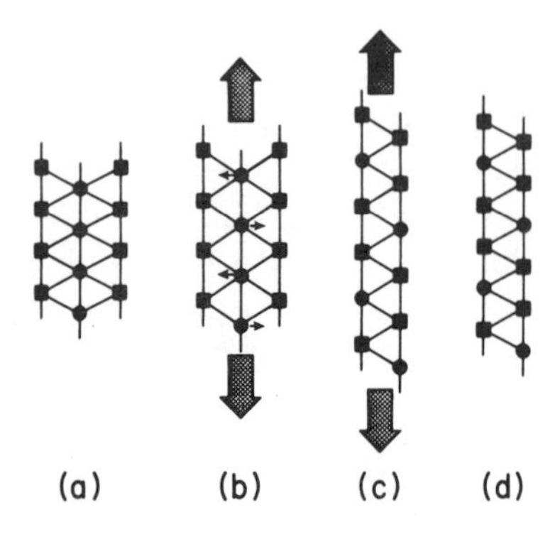

Figure 12.1. Elastic and plastic deformations explained by the dislocation of submicroscopic particles. The sequence (*a*) (*b*) (*a*)—without the tiny arrows—illustrates elastic deformation, the sequence (*a*) (*b*) (*c*) (*d*) plastic deformation.

Viscoelastic Nature of Wood

Elastic materials are by definition not permanently deformable, their particles do not slip when the piece is stressed. By contrast, in liquids and viscous substances such as honey, the slightest force pushes loosely connected particles into new neighborhoods, causing flow. Wood falls between these two extremes and is considered *viscoelastic,* a possibly misleading term; *elasto-viscous* would be more descriptive since at ambient temperature wood behaves predominantly in an elastic fashion.

The dominance of elasticity shows in the longtime behavior of bent pieces, which in adsorption-drying cycles gradually recover and approach their original posture. Wood has an *elastic memory:* its original rods between particles never break, they only overstretch, become thin, and pull much less than the rods between new neighbors. In adsorption-drying cycles, original rods gradually regain dominance and reverse the slippage.

Slippage also explains why wood has a proportional limit (Section 4.1) and why it creeps or deforms under constant stress (Section 4.3). Before particles slip, certain stress increments cause certain increases in strain, the two being proportional. Above the proportional limit, or under enduring load, slippage upsets the proportionality through its resulting additional strain. Actually the proportional limit depends on the rate of loading, being relatively low when the load increases slowly.

The force required to hold a bend diminishes as time passes, the material resists the imposed deformation less and less, the internal stress is relieved or relaxes. *Stress relaxation* is defined as *stress decrease under constant strain.* Slippage explains this phenomenon too, since no force at all is required to hold the piece in Fig. 12.1*d* at the length of the piece in Fig. 12.1*b*. Stress relaxes due to slipping particles or *internal creep.*

12.2 BENDING CURVATURE

The preceding section showed how to achieve permanent bends, while this one concerns the degree of forming and explains how to produce relatively sharp bends. To what extent heat and moisture contribute to sharp bends depends on increased ultimate strains, as shown in Fig. 12.2. At ambient temperature dry wood stretches about 1%, while steamed it endures tensile strains up to 2% and compressions of much higher magnitude.

These deformations or strains are still inadequate for many products, such as the sharply bent parts of chairs. To increase bends, the industry applies a technique invented already early in the 19th century by the cabinet-maker M. Thonet, who observed that wood fails at the convex stretched side, while the fibers at the concave compression side may buckle but remain fast. Pressure against the ends of the piece, mostly provided by a bending strap (Fig. 12.3), restricts stretching at the expense of compression at the concave side, thus allowing much sharper bends.

The large compressions involve microscopic failures in the form of crimpy and buckled fibers, which affect strength to an extent tolerable for most purposes. Deformations of this kind are essentially plastic in nature and contribute to the permanence of the bend. It is therefore not necessary to bend farther than the desired shape, except where adsorption-drying cycles later, in service, cause some gradual recovery in sticks which are not fixed as parts of larger products, such as chairs.

The smallest possible bending radius r depends on stick thickness t and on the ultimate difference Δs in strain between concave and convex side:

$$r = \frac{t}{\Delta s} \qquad \frac{r}{t} = \frac{1}{\Delta s}$$

For a stick that can be stretched and compressed 1% each, for example, Δs is 0.02 and the bending radius may not be less than 50 times the thickness. With a bending strap that compressed 9% and allows 1% stretching, the ratio of *bending radius to thickness* drops to a value of $1/(0.09 + 0.01)$ or 10. In species of wood well suited for bending, named below, the average safe ratio exceeds 30 in the dry state at ambient temperature, 13 steamed, and 3 steamed with strap. Some researchers have found a limit of 60 for dry

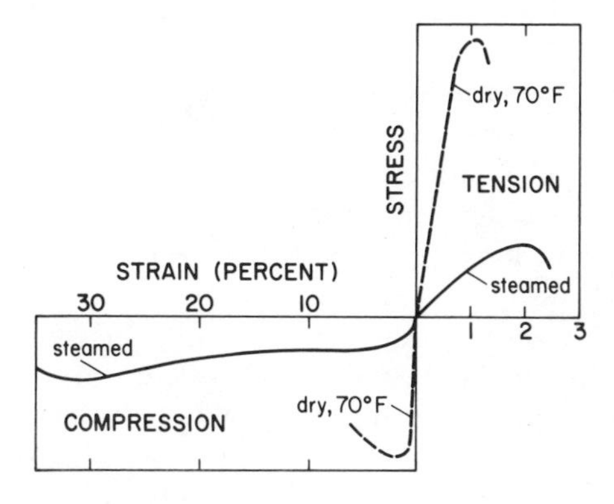

Figure 12.2. Effect of steaming on compressibility and stretchability of temperate zone hardwood. Note the different scales for compression and tension.

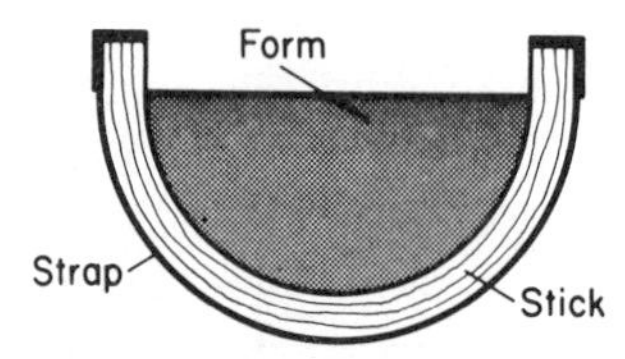

Figure 12.3. Bending with strap.

wood, while in some cases, in the steamed state with strap, the radius could be even less than the thickness.

Heavy remperate zone hardwoods bend farther and retain a higher percentage of the deformation than do softwoods and tropical hardwoods. This difference is attributed to lignin, the constituent that embrittles wood and is relatively low in temperate zone hardwoods (Section 2.1). The force required to steam-bend dense wood of this group exceeds only slightly the force needed to steam-bend common softwoods, because steaming softens the hardwoods more. Figure 12.2 illustrates the strong softening effect in the low *stress-to-strain* ratio when steamed, compared with dry wood at 70°F. Since in bending with straps most of the deformation consists of compression, the flat compression curve indicates that steamed hardwood bends easily.

Opinions vary as to which species among heavy temperate zone hardwoods bend best. Some manufacturers of bent products prefer the tough hickory, a choice which makes sense since *toughness* means *high ultimate strain;* others claim that elms, oaks, beech, birch, maple, and gum are just as good.

12.3 PLASTICIZING WITH AMMONIA

Water plasticizes wood by swelling the cell wall substance and squeezing the particles apart into positions of easy slippage. However since moisture enters only between aggregates of cell wall molecules, the effect is limited. Anhydrous ammonia (NH_3, not to be confused with household ammonia which is a water solution of NH_3) penetrates these aggregates as well, breaking hydrogen bonds between molecules and permitting the molecules to slide past one another under stress. The chemical facilitates slippage, it plasticizes much more than water, and renders veneer as flexible as damp leather.

The ammonia breaks all hydrogen bonds only when highly concentrated in the cell walls. In turn this high concentration in walls requires high concentration in cell cavities and around the entire piece, for the following reasons. Ammonia migrates by molecular motion, driven by concentration gradients, first into the cavities and then into the walls, the molecules continuing to diffuse from site to site and also back out, just as moisture does. In the eventual state of equilibrium, the number of molecules diffusing out equals the number of molecules diffusing in. The required high concentration around the wood is achieved either through condensing, by cooling below the boiling point of −28°F (−33°C), or more speedily through compression to about 145 psi (150 psi causes condensation at room temperature).

For bending, expose the wood to ammonia until its concentration in the cell walls reaches the required high level. Upon removal, bend the piece as long as it retains spectacular flexibility. The ammonia rapidly evaporates, diffusing out of the cell wall, in which new bonds form at new positions of molecules and particles, with the effect that the wood regains its original stiffness and strength within an hour while retaining the bend. Very little if any clamping is required, since holding the bend for a few minutes sets the shape. In service, ammonia-bent wood remains stable even in hot water and during wetting-drying cycles.

Ammonia treatment however causes a permanent change in width and thickness. The penetrating chemical swells cell walls, with the effect of increasing wood dimensions and reducing cell cavity space. During drying, the gas first migrates out of the cell wall layer close to the cell cavity. The inner shell therefore regains rigidity while occupying its original cavity space, and forcing the other cell wall layers to shrink towards the shell, with the result that cell walls crimple and wood volume decreases by 10% to 40%, compared with volume before treatment.

Otherwise, with regard to plasticity and long-range influence on wood properties, anhydrous ammonia leaves nothing to be desired, since it allows *bending radius-to-thickness* ratios as small as 4 without a bending strap. The severe drawbacks, however, are the required cooling or compression of the gas and the relatively long treatment times. Veneer 1/8 in. thick must remain in ammonia for about 1/2 hr. Thicker wood calls for much longer treatment, since penetration controls the rate of plasticizing; treatment of lumber is therefore too expensive. Typical ammonia-bent products are veneer spirals and helics, as parts of exotic lamps and other novelties. The bulk of solid wood bending is still carried out in the traditional way after steaming or boiling.

12.4. TRANSVERSE DEFORMATIONS

Plastic transverse deformations occur in two forms: as evident densification, and as *sets* (plastic deformations, Section 3.9), which are too small to be visible as such, but interfere with woodworking and devalue some final wood products. Compressions of wood-base panels amount to densification, however small, and are treated here accordingly.

Densified Wood and Compressions of Wood-Base Panels

Moist heat makes cell walls so pliant that under transverse pressure they crumple and fold without breaking, bulging into cell cavities which collapse and finally disappear (Fig. 3.10). Wood density increases accordingly and approaches the density of the cell wall substance. Densified wood, coined *staypak,* is as strong as natural wood of the same density would be. The densification renders species as light as aspen stronger than any natural wood.

During pressing, heat activates lignin and other wood constituents as binders, which hold the walls of the flattened collapsed cells together, so that under fluctuations of humidity staypak will remain compressed. Only boiling in water causes recovery. Europeans use staypak from beech for loom shuttles and other highly stressed items. In North America no industrial applications have been reported.

Compression of plywood, particleboard, and hardboard never reach the degree of staypak, even though the panels are pressed at high temperatures. The pressures are too low to flatten the cells. Set in plywood has a magnitude of 5%. Many particleboard and hardboard panels are compressed more, but opposing walls of cell cavities still remain apart, so that the sets *spring back* and make the panels thicker when they swell and shrink, the degree of springback depending on added binders and pressing conditions.

Sets from Restraint of Swelling and Shrinking

This form of set occurs in the handles of axes, hammers, and other tools in which steel encases wood. Manufacturers fit the dry handles into the steel head, leaving no allowance of course for swelling. In use, adsorbed moisture generates swelling pressure against the unyielding steel, and the handle is compressed in comparison with free-swelling wood. Small moisture gains cause elastic compressions which disappear again when the wood dries, but large adsorption results in compression set. During redrying, wood which is in set shrinks almost like normal wood, so that on return to its original moisture content the handle is thinner than originally and fits loosely. Keep tools therefore out of water and rain, or impregnate the handle with oil, or coat the end face with a water repellent, or seal it with paint.

In tight wood finish floors the set resulting from moisture adsorption causes open joints to appear between redried boards. Dirt fills the gaps and restricts swelling when the wood readsorbs moisture. In each adsorption-drying cycle the gaps become slightly wider so that very old wood floors show distinct streaks of dirt between the boards. Fortunately the streaks develop slowly and only under large moisture fluctuations. Modern hardwood strip flooring, which is kept clean and sealed, remains tight under normal moisture conditions.

It has been surmised that the ancient Egyptians and West Indians utilized the swelling pressure of wood to split rocks for their pyramids. The pressures can indeed be tremendous, but splitting rocks with wood is a tedious time-consuming process. The swelling pressure equals the pressure required to compress wood, which is small in the range of low compression, as illustrated in the *perpendicular test* of Fig. 4.3. High pressure develops only after all cell cavities have collapsed in numerous swelling-shrinking cycles under swelling restraint. The ancients had to drive dry wooden wedges into a rock cleft, wet the wedges, let them dry, drive the shrunk wedges deeper, rewet, and so on until internal swelling has squeezed the entire cell cavity space out of the wood. In this way Egyptians and Indians could generate pressure in the magnitude of 30,000 psi.

Restraint of shrinkage has consequences similar to restraint of swelling and causes permanent stretching, though of low magnitude. When wood dries and shrinks at very different rates near the surface and in the center, the resulting stresses generate sets of the tension and compression types, known as casehardening, mentioned in the Drying Section (3.9).

Significance of Transverse Plasticity

As a raw material wood competes with metals and plastics, over which it has the natural advantage of growing as needed, while metals and plastics must be manufactured from scarce and limited deposits in the earth. Wood is therefore cheaper per unit of strength as well as per unit of volume; only foamed plastics in which air comprises most of the space cost less. Compared with wood, a short solid pillar of certain load-carrying capacity costs two times more in steel, three times more in the cheapest plastic, and many times more in aluminum.

Nevertheless lack of plasticity, a crucial property in these days of mass production, impairs wood's competitive position. In manufacturing a large series of items from plastic material, it pays to make an expensive form of the desired shape for pressing or casting the item quickly without waste. In this way the industry manufactures aluminum dishes and plastic spoons—to name just two examples—for pennies per dozen. By contrast, wooden dishes and spoons must be shaped laboriously and wastefully by chipping material away.

In other products wood remains competitive but has lost ground. Wooden door jambs and casings, for example, have to be machined and installed in three parts, two casings and one jamb, while in metals and plastics the combination is pressed as one piece. The industry shapes U-studs from sheet metal for non-load bearing walls, which sell at below the cost of lumber studs. As floor support, rectangular lumber joists must compete with steel I-beams, which have most of the material in the flanges where the stresses are highest. Unfortunately, unlike hot steel, trunks of trees cannot be rolled or squeezed out in efficient I- and U-forms. Wood is limited in plasticity; it can be compressed but not stretched to significant degrees, and has to be molded by chipping.

13

Gluing of Wood

Since wood does not lend itself to plastic forming, it must be shaped either by cutting off material, or by connecting pieces in desired configurations. Connecting pieces by melting and fusing (as in welding metals and heat-sealing plastics) is impossible because heat destroys wood. This leaves us with only mechanical fasteners and glue-joining. Mechanical fasteners such as nails and screws connect only at points, a merest fraction of the surface, and their hold is relatively small. Gluing provides the strongest and most elegant, unobtrusive connections; it finds wide use in furniture and is essential in the manufacture of plywood, glulams, and particleboard.

The term *glue* originally stood for *bonding materials of natural origin*, but in woodworking now all bonding agents are called *glue*. Some wood technologists and wood researchers use the more comprehensive terms *adhesive* and *adhesive bonding*. I prefer the terms *adhesive* for theoretical and *glue* for practical matters.

13.1 PRINCIPLES OF ADHESIVE BONDING

Smooth-planed or sanded wood looks anything but smooth under submicroscopic magnification. The surfaces resemble a rough terrain with deep valleys, steep slopes, and sharp mountains, which in joints touch each other at only few points (Fig. 13.1). To achieve more intimate contact, the space between the contours must be filled with adhesive that penetrates all crevices, voids, and depressions. For that, the adhesive must first wet the surfaces and should therefore be fluid; later it must solidify (*harden, set*) for transmission of forces. The hardening process determines the appropriate gluing procedure and differs among the various adhesive types, which are here described.

Solvent-Loss Type These adhesives are solutions or stable emulsions, such as *white glue*, and consist of a binder dissolved or dispersed in water or another solvent, whose adsorption by wood causes the hardening. Their type is well-known on the backs of postage stamps, where licking redissolves the glue and makes it fluid enough to be pressed into the envelope's pores.

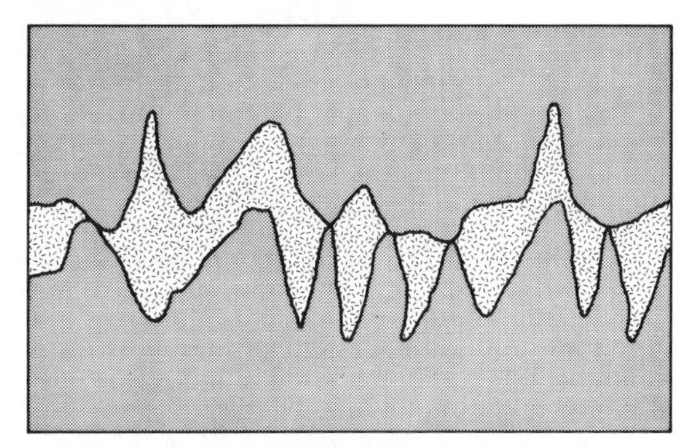

Figure 13.1. Theoretical submicroscopic adhesive joint.

The bond develops rapidly as paper adsorbs the moisture. Should stamp and envelope later be rewetted, the glue would reabsorb moisture and revert once again to the fluid state, in which it ceases bonding. Solvent-loss adhesives generally set reversibly, and are therefore not durable under exposure to moisture if water happens to be the solvent.

Chemically Curing Type Adhesives in this category set or *cure* by chemical reaction. Most consist of low-molecular-weight dissolved polymers that combine or polymerize to high-molecular-weight insoluble polymers. Strictly speaking, few such adhesives cure by polymerization, most undergo polycondensation, that is, polymer formation through departure of molecular water or of some other simple substance, which has to migrate out of the glueline.

In *cold-setting* adhesives of this type an added *hardener* promotes polymerization, while in the *hot-setting* kind elevated temperature triggers the setting. Hardeners are consumed in the curing process and participate therefore in significant amounts, whereas *catalysts*—a kind of hardener in a broad sense—speed up the reaction without being consumed, so that small quantities suffice.

Heat accelerates the reaction in any case. Some adhesives which set at room temperature within the hour will cure in a few seconds above 250°F. Chemically curing adhesives are therefore called *thermosetting* adhesives, particularly because some harden only when heated.

Polymerization and polycondensation are irreversible chemical reactions, a fact that contributes to durability of their bonds. Heartwood extractives which diffuse into the glueline may disturb the chemical curing process. This is why heartwood is considered more difficult to bond than sapwood.

Hot-Melt Type The adhesive is furnished in solid form and melts when heated. After spreading the liquid, users rapidly join the mating pieces before the adhesive can bond by solidification, which happens the instant heat dissipates out of the glueline. This high speed is the reason why hot-melts dominate in continuous edge banding operations.

Hot-melts belong to *thermoplastic* adhesives, meaning adhesives capable of being softened repeatedly by heat. White glue and animal glue have thermoplasticity too, but do not fall into the *hot-melt* category, since white glue hardens by migration of solvent out of the glueline, animal glue by solvent migration and cooling.

Pressure-Sensitive, Contact Type After spreading, these adhesives quickly attain a consistency inbetween solid and liquid; being apparently dry and with no ability to flow, they require strong pressure to make them penetrate the surface microcontours. The thick consistency and limited flow gives this type considerable tackiness and the ability to bond instantaneously upon contact.

Rubber cement for patching innertubes is a well-known contact type, except that its curing also involves loss of solvent. After it is spread on innertube and patch, the solvent must be allowed to evaporate, since the nonporous rubber cannot adsorb it. After solvent loss, the cement reaches a pressure-sensitive state and bonds when pressed.

The wood industry uses rubber-base contact type adhesives for bonding overlays to panel surfaces in basically the same fashion. The bonds hold adequately right away, reaching their final strength (near that of other glues) later on, when small amounts of solvent diffuse into the wood. Adhesives on plastic contact-foil, masking tape, cellophane tape, self-sticking labels and the like are of a solvent-free pressure-sensitive type.

Combined Types Generally glues for wood set by more than one type of hardening; most lose solvents in addition to another process. The thermosetting synthetic resins, for example, cure mainly chemically, but solvents and polycondensation water have to diffuse into wood in time with the chemical reaction. When the solvent escapes too early, some adhesives will fail to harden at all, while tardy solvent loss may cause *crazing*, that is, breaking up of the glue layer with many tiny cracks. The chemical reaction and loss of solvent coincide almost automatically when heat triggers both.

13.2 GLUING PROCEDURE

For adhesive bonding, condition the wood to the proper moisture content, prepare the surfaces, spread glue, press the mating pieces together, let the glue set, and condition the bonded product. These steps are discussed here in their natural order.

Conditioning

Wood moisture influences gluing through migration of solvents out of the glueline, through wood swelling and shrinking during glue setting, and through shrinking, swelling, and warp after setting. Excessive moisture retards the adsorption of solvents, and causes wood to dry and shrink during as well as after pressing, which matters because the two pieces tend to shrink by different amounts and to slide relative to each other. Drying after pressing causes warp. Pieces that contain too little moisture adsorb the solvent too fast, so that the glue loses its ability to flow before pressing, or else it may not harden at all. Very dry glued products adsorb moisture from the air, swell, and warp just as moist products warp when drying.

Most glues are formulated for 8 to 12% MC, the moisture content in equilibrium with normal use conditions. For different initial and final moisture contents the gluing operation has to be modified. If the wood is wet, for example, spread low-solvent glue, leave plenty of time for absorption of wood moisture, and allow moisture to evaporate before pressing. (Note: the sucking up of moisture by glue is **absorption**, a *taking in*, whereas dry wood **adsorbs** moisture on its large internal surface, as explained in Section 3.5).

Wood Surface Preparation

Zigzag contours interlock and have much greater actual surface than do planed pieces. It is also known that glass surfaces are too smooth for adhesive bonding. These facts might lead to the erroneous conclusion that strong bonds require a rough wood surface. Planed wood **has** sufficient roughness, however, while unsurfaced lumber is blanketed with frayed and torn-up fibers which have little connection to the solid wood underneath and separate easily, causing the bonded product to fail at this fiber-wood interface. Wood surfaces cannot be too smooth for gluing. Some hardboard, particleboard, and plywood surfaces that were *closed* during hot-pressing make the exception for some glues and do require slight sanding for better adhesion. Smooth but uneven surfaces give locally thick, conspicuous glue lines which distract from the appearance of the product and also have low strength, as explained below under *glue spread*.

Pieces that were machined many months before gluing may give adequate bonds, but freshly planed or sanded surfaces hold better. The latter are clean, not besmeared by exuded resin or oil, and free of warp from drying or moisture adsorption. It also seems that machining temporarily *activates* wood for bonding, as if newly exposed molecules look out eagerly for affiliates. Gluability decreases with storage time so much that one-year-old surfaces may hold one-half as well as fresh surfaces. If more than one side of a piece has to be glued, the precautious craftsman machines each face shortly before the particular glue spread.

Glue Spread

From Fig. 13.1 may be surmised how *thick* or viscous the glue should be, how much glue to spread, whether spreading on one surface only will do, and how much time should pass from spread to assembly and to pressing. The glue is not intended to separate the two pieces, it should rather interlock their submicroscopically rough surfaces by filling the voids, so that cohesion within the glue matters little and the joint's strength actually exceeds that of the glue.

A thick glueline represents a distinct link which is likely to be weaker than either one of the two pieces. Once the microvoids are filled, bond strength decreases as glueline thickness increases, the reason being that glue solvent swells the wood and causes stress in the bond. Besides that, thick gluelines tend to become porous and to craze (Section 13.1) from loss of the solvent during setting and from the resulting shrinking of the glue itself.

The actual spread of glue may be controlled through choice of glue viscosity and through the method of spreading, following the principle *spread thick glue thin*. Large consumers mix the glue as thick as possible within the limits of the small spread desired. For roller spreading, the viscosity of fresh honey serves as a rough guide, while for brushing the glue has to be thinner. The lower limit of viscosity depends on the gluing pressure, which if excessive forces thin glue into porous surfaces, leaving nothing for filling the voids and causing *starved* joints. Porous cross-grain and end-grain surfaces require relatively thick glue; alternatively they are sealed with a preliminary coat that may contain a *filler* and which dries before the bonding coat is spread on.

Coating of one surface serves the purpose, provided the spread is sufficiently large and fluid to flow into the voids of the mating surface. Double-spreading becomes necessary if much *open* time passes between spread and assembly. With glues that set quickly after hardener is added, instead of spreading the mix it may be practical to spread the hardener on the mating surface, so that it diffuses into the resin to react at the right time, during pressing.

If during the open time too much solvent evaporates, the glue *skins* and does not flow when pressed. However, some gluing tasks not only permit long open time, they even require it to keep solvent out of the wood and to reduce glue shrinkage during pressing. The industry waits as long as possible before assembly and before pressing, to save press time. In veneer about 1/8 in. thick and less, most of the glue solvent has to evaporate before assembly because the thin layer cannot adsorb it all and the thin glue would bleed out through thin face veneer. When bonding plywood to plywood, the panel gluelines retard solvent migration into deeper layers and may also disturb the setting. If it is desired that glue moisture be retained, in order to raise the moisture content of the glued product to the moisture level in use, store the assembled pieces before pressing. Denser species require relatively long open times and long *assembly* times because they adsorb slowly.

Some adhesives are marketed in the form of dry films. Inserted between the layers to be bonded, they melt in the hot-press and flow under pressure into micro-voids of the two surfaces, hardening by heat-triggered chemical reaction. Such *glue by the roll* is bleed-through-proof even under thin cross-grained veneer.

Pressing, Setting, and Final Conditioning

Gluing pressure forces glue into surface voids, irons air bubbles out of the glueline, squeezes excessive glue out also, and prevents the two pieces sliding relative to each other due to swelling and shrinking. Choose the pressure accordingly. Glue of high viscosity that flows little, and dense wood species which develop high swelling forces, both require heavy pressure. Dense species need high pressure also to bring uneven surfaces into contact. The required high pressure is one reason why dense wood is relatively difficult to glue.

Optimal pressures, achieved in hydraulic and pneumatic presses, range from 100 to 250 psi, which is at ambient temperature far below the fiber stress at proportional limit for all wood species. Heat however lowers the proportional limit so much that hot-pressed wood frequently retains some compression. Screw clamps, whose force rarely exceeds 250 lb, have to be very close together to generate more than 100 psi; their pressure drops due to softening of moisture-adsorbing wood and they should be readjusted after the first 15 min. Glue squeezing out on exposed edges of gluelines gives some indication of sufficient pressure. For glues with good *gap-filling* characteristics (Section 13.3), even nailing pressure gives adequate bonds, while small well-planed surfaces require no pressing at all.

Most glues used in small workshops harden at ambient temperature at a rate allowing release of pressure within the hour. Factories cannot wait that long and so use particularly formulated glues in hot-presses, or heat by means of microwaves to above 250°F to set the glue in a few seconds. In hot-presses it takes much longer than that for heat to flow from the press platen through the insulating wood to the glueline.

According to one rule of thumb, the pressure can be released as soon as the *squeeze* is hard. At time of release the bond should be strong enough for handling and to withstand weakening under the wood's swelling and shrinking. The curing process continues after hot-pressing, coming to an end hours or days or weeks later. Last traces of solvents diffuse out of the glueline only after months, although frequently without strengthening the bond, because the solvent loss in fact embrittles the glue.

13.3 STRENGTH OF ADHESIVE BONDS

You can test joints simply by forcing a flat screwdriver into the glueline and twisting it to pry the two pieces apart. Good side-grain bonds separate mainly in wood, not in the glueline, an indication that in tension the bond is stronger than wood (perpendicular to fiber direction). Wood failure serves as a criterion of the bond also in a quick shear test, striking the blow of a hammer against one of the two pieces in the direction of the glueline. Testing machines evaluate bonds more measuredly by pulling lap joints apart (Fig. 13.2) and indicating the required force, for calculating the shear strength in psi.

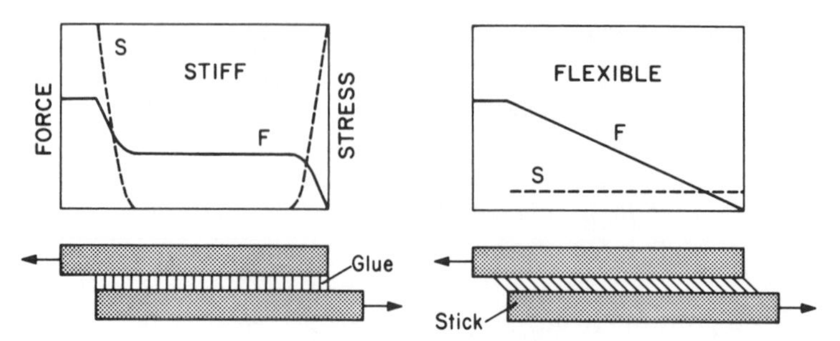

Figure 13.2. Distribution of shear stress S and tensile force F in lap joints with stiff and flexible glue.

Effect of Adhesive Flexibility

Shear strength and percentage of wood failure do not necessarily correlate because adhesive flexibility affects shear strength and wood failure differently, as illustrated in Fig. 13.2, in which the thick glueline transfers tensile force from one stick to the other through shear stress. The stiff adhesive, which does not deform at all under stress, transfers the force in two points at each end of the glueline where the shear stress peaks, with both sticks sharing the tensile force equally where they overlap. By contrast, a flexible adhesive joint transfers the force over its entire length, to the effect of there being small and uniform shear stress in the whole joint and gradually diminishing tension in each stick.

The stiff joint naturally fails first near the ends at the two points of stress concentration, either in wood or in glue. Then the points of stress shift centerward, concentrating at the ends of the remaining intact glueline, where the bond fails again. In this way stress and failure rapidly spread from point to point along the whole glueline, independent of glueline length. Flexible adhesive transfers load over the entire length and resists higher loads, possibly without any wood failure at all.

Adhesive rigidity explains why shear strength of lap joints depends on thickness of the wood. A certain pulling force causes high tensile stress and considerable stretching in thin wood layers (as in Fig. 13.2). Rigid adhesive cannot stretch along with the wood and therefore transmits much force near the end of the overlap. Again a stress peak appears and causes early failure. By contrast, in thick wood layers the same force causes only small stress and little stretching which the adhesive can endure, with the effect of transmitting the force evenly over the entire length of overlap, and achieving high strength in the glueline as a whole.

Flexible, ductile adhesives are *gap-filling*, meaning they are *little affected by excessive glueline thickness*. They give bonds of adequate strength between members that lack close contact, such as rough lumber or wood glued under low pressure.

End-Grain Bonds

End-grain bonds (Fig. 13.3A) should be as strong as wood in fiber direction, that is, many times stronger than side grain bonds, while in reality end grain bonds are much weaker, for several reasons. The rough surfaces, blanketed with torn up fibers, lack close contact. Even if smoothed with very sharp tools or fine sandpaper, good contact is still restricted to a fraction of the area where cell walls of one piece face cell walls of the mating piece rather than cell cavities (in wood of specific gravity 0.5, the cell wall's specific gravity being 1.5, the wall-to-wall contact area is only 0.5/1.5 of 0.5/1.5, or 1/9 of the joint area.) Finally, glue sags away into cell cavities instead of staying in the glueline.

Scarf joints (Fig. 13.3B) have relatively large, nearly lateral surfaces and are therefore much stronger than *butt joints*. Since wood is approximately 10 times stronger in tensile stress than in shear stress, scarf joint surfaces should be at least 10 times larger than the cross section of the pieces, to reach the wood's longitudinal strength. The tensile strength of *finger joints* (Fig. 13.3C and D) falls between that of butt joints and scarf joints, depending on the angle of the fingers and on the size of butt end areas at the finger tips (Fig. 13.4). In bending, finger joints fare better, of course.

13.4 DURABILITY OF ADHESIVE BONDS

Durability is closely related to adhesive strength, since it concerns the effect of exposure on strength. Glue joints can last for ever if nothing damages them; they may be weakened and destroyed not only by fire, living organisms,

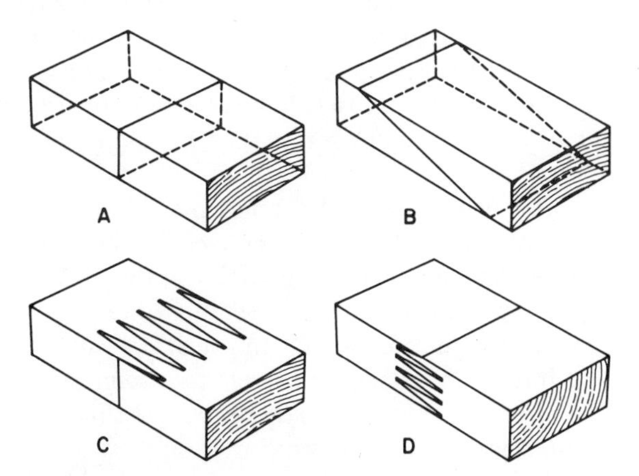

Figure 13.3. Types of end-to-end grain joints for lumber. *A*, Butt joint. *B*, Scarf joint. *C*, Vertical finger joint, the most common type for structural purposes. *D*, Horizontal finger joint. (Photograph courtesy of U.S. Forest Products Laboratory.)

Figure 13.4 End-to-end grain finger joint. (Photograph courtesy of National Casein Company.)

and mechanical force, but also by moisture as such and by changing moisture contents. In fact moisture is by far the most important factor in glue bond durability.

Moisture influences the bonds in the first place through wood swelling and shrinking, which as such do no harm as long as the mating pieces change size by equal amounts in the glueline plane. But as a rule the two pieces differ in their swelling and shrinking, especially if one is flat-grained and the other edge-grained or one dense and the other light, and most of all in crossbanded joints. The bond compels the two pieces to swell and shrink identically in one compromised movement, with the more stable piece restraining the other. The resulting stress on the bond naturally increases as thickness of the bonded layers increases; bonds between boards, for example, have to resist much stronger swelling forces than bonds between veneers, and are therefore less durable.

Some stress results from the bond forcing the adhesive to expand and contract in the glueline plane along with the wood. Glues of natural origin and white glue themselves swell, resembling wood in this regard, and seem to have an advantage over nonhygroscopic synthetic adhesives. Their swelling however is not exactly like that of wood, and it also involves softening and loss of strength; in humid air some glues revert to a semi-fluid state and fail under minor stress. Thermosetting synthetic resins resist moisture better, and have attained for this reason a dominant role.

Manufacturers designate glues as *moisture resistant, water resistant, waterproof*, or *boilproof*, terms which in themselves are meaningless, for two reasons. First, *waterproof*, for example, means *not affected by water*, which for glue joints is impossible since moisture weakens the wood in any case and promotes wood failure. Second, it is the effect of moisture on the bond and not on the glue that counts. Bonds may last in water in spite of having been made with a water-soluble resin.

Wood technologists characterize glue with the quality of the bond as defined in the plywood standards (Section 6.2). *Interior glue* is suited for Interior type plywood, *exterior* or *waterproof glue* has the quality required for Exterior type plywood. The plywood standards avoid the terms *moisture resistant, water resistant* and *boilproof*. Still some people will designate bonds of Interior type plywood as *moisture resistant,* somewhat more durable bonds as *water resistant*, and Exterior type plywood bonds as *boilproof*, referring to the boiling test which samples have to pass.

Fungi and bacteria prefer natural glue over lignified wood cells as food and concentrate on such gluelines, destroying the moist joint unless a preservative has been added to the glue or incorporated in the wood before gluing. Synthetic adhesives are safe, since so far neither fungi nor bacteria have learned to decompose them.

Under exposure to heat and fire, the performance of glued products depends mainly on the bond. When the joints of plywood and laminated veneer lumber fail, the separated laminates burn like kindling and literally burst into flames. Glue bonds qualify as *fire resistant* if they hold when wood chars, that is, if the glued product decomposes without delamination, as required for Exterior type *Construction and Industrial Plywood* under exposure to flames.

13.5 KINDS OF GLUES

The number of brands and varieties of woodworking glues on the market in North America goes into the hundreds and thousands. Many brands contain the same binder but differ in added *fortifiers, fillers, extenders*, and solvents, and in particular processing methods.

Fortifiers represent binders that are more durable than the basic binder and improve durability. *Fillers*, though nonadhesive substances such as flour from walnut shells, still influence glue properties, particularly the viscosity; their job is to fill pores and enlarge glue volume. *Extenders* like wheat flour have some adhesive action and their purpose is to reduce the amount of binder required per unit area.

Buyers choose glue mainly by price, durability under exposure to moisture, and readiness for use. Price increases as resistance of the bond to moisture improves; in other words, durability must be paid for. Below I shall discuss the various kinds of glues in the approximate order of rising durability. If a glue appears to be expensive in spite of low durability, it probably comes ready to use.

Large consumers buy glue either in the form of powder to be mixed with solvent and other agents such as hardener, or as a liquid to which hardener may have to be added. Small workshops prefer prepared glue that sets at room temperature. Powders generally have unlimited shelf life, whereas liquid hot-setting resins undergo slow polymerization and therefore keep only a few months even in cool storage. This polymerization process naturally limits the *pot life*, that is, the time period over which ready glue can be used. Growth of fungi and bacteria shortens the pot life of water-based glues of natural origin.

Glues of Natural Origin

The least durable and cheapest glues are made of starch. Marketed in the form of flakes and powder to be dissolved in water, starch glue is sufficiently strong for wallpaper but not for wood. The wood industry uses this kind only as an extender.

Cabinetmakers used to bond exclusively with *animal glue*, the broken down protein from extracts of hides and bones, which has sufficient strength for furniture and is unsurpassed in appearance, since the gelatine-like adhesive cannot be seen in thin gluelines. On the other hand animal glue is inconvenient to work with. Users soak and heat purchased beads or bars before spreading the melted glue on preheated wood, on which the glue rapidly gels through cooling and loss of water. Bonds of simple animal glue do not hold under moist conditions.

Protein glues made from soybeans, dried blood of cattle, and milk offer more moisture resistance. *Soybean glue* is still not sufficiently durable for *Hardwood and Decorative Plywood*, but it qualifies for *Construction and Industrial Plywood* of the Interior type, particularly if fortified with blood-base binder—a binder that resists moisture nearly as obdurately as bloodstains in fabrics resist washing. *Casein glue*, derived from milk with calcium hydroxide added, is of the chemically reactive type, sets cold, and tops glues of natural origin in water resistance. At room temperature, casein glue bonds hold if wood moisture does not exceed 18% for long periods of time. The conservative glulam standard allows casein bonds where the moisture content in service does not exceed 16%.

White Glue

White glue is a milk-like stable aqueous emulsion of dispersed polyvinylacetate droplets, which resists water least, among the common synthetic adhesives, ranking inbetween animal glue and casein glue in this regard. But where no liquid water penetrates the wood, white glue is durable enough for many products. Hobbyists prefer white glue because it is ready to use, keeps for years, sets at room temperature simply by loss of water, and gives nearly invisible translucent gluelines as the white color clears up during setting. The glue retains great flexibility, which helps resist load of short duration, but which under long lasting load leads to cold flow (creep). Its flexibility and thermoplasticity also clog up sandpaper.

White glue costs more than other common adhesives, but the price has little significance for hobbyists who use small amounts. The name *white glue* does not appear on the label but users and dealers prefer this descriptive term. More recently similar adhesives, not polyvinylacetate-based, have appeared purporting to have advantages over polyvinylacetate glue; people call them *yellow glue* (or also sometimes *white glue*).

Thermosetting Synthetic Adhesives

Most plywood, particleboard, and furniture are bonded with hot-setting *ureaformaldehyde resin*, the common adhesive for moisture resistant, Interior-type bonds. Some glue of this kind is formulated for cold-setting in the home workshop. In warm humid surroundings, urea resin bonds do not hold well, separating in boiling tests after a few hours. The cured resin forms inconspicuous gluelines of light or slightly tan color, but it is brittle and therefore gives weak bonds in thick gluelines. Urea dulls tools faster than other thermosetting synthetic resins and faster than casein glue containing calcium (other natural glues and white glue have no more effect on tools than wood has).

Phenol-formaldehyde resin bonds surpass those of urea in durability. This resin excells also in toughness and flexibility and is the preferred adhesive for waterproof, Exterior type plywood. It furthermore ranks highest in strength, also in dry joints, above urea, casein and white glue, soybean and blood glue, and animal glue at the lower end of the series of common woodworking glues. Phenol glues are typically formulated for curing near 300°F. For waterproof bonds with cold-setting adhesives, manufacturers of glulams and hobbyists use the more reactive and more expensive *resorcinol resin*, which resembles phenol chemically and physically; it polymerizes after mixing with a hardener. Phenol and resorcinol resins are dark red before and after curing. In gluelines this color may speak for waterproof bonds, but not all dark red glues are waterproof, nor are all waterproof joints red. *Melamine resin* is colorless and approaches phenol in durability.

Elastomeric Adhesive

These relatively new adhesives rank fairly high in moisture resistance; their basic ingredient is rubber or a rubbery substance. The group is called *mastic adhesives, mastic* meaning *pasty* in this context, a state that allows marketing in cartridges and application with caulking guns, squeezing beads out of the cartridge nozzle directly on the wood.

Elastomers are defined as *substances which can be stretched repeatedly to at least twice their original length and upon release of the force return to original length*. Adhesives with this extremely high degree of flexibility have therefore excellent gap-filling characteristics for rough wood surfaces, low gluing pressure, and the thick gluelines which result. Elastomers are used in construction in combination with nailing, for example to bond subflooring to floor joists and panels to walls, the spread being in strips rather than in whole areas of contact. Use in construction gave rise to calling the group *construction adhesives*, a term that should not imply strength, since for high strength the adhesive is too viscous and is pressed too little.

Elastomeric adhesives are of the pressure-sensitive, contact type and bond immediately, hardening little, mostly by diffusion of a dispersed component into wood, and/or by chemical reaction catalyzed by moisture or oxygen. Most brands retain some resiliency, which is advantageous under stress of short duration but permits creep under loads of long duration. Caulking substances (sealants) resemble mastic adhesives in many respects but are formulated for more resiliency.

Epoxy Resin

The chemically curing *epoxy resin* is considered stronger than common adhesives. Some people use it for wood, although it costs a good deal and has to be mixed with hardener shortly before each application. If for some reason the glueline must be thick, epoxy may provide stronger bonds than common glues, because it does not shrink during hardening and cures to a strong solid. But normally epoxy resin offers no advantage in gluing wood, since glue bond tests of many woodworking glues give high percentages of wood failure, leaving nothing to be desired of the glues themselves, at least not for side-grain bonds.

For bonding wood to nonporous materials, epoxy has the advantages of containing no solvent and not splitting off any compound during polymerization, which explains the absence of shrinking. Epoxy resin holds fast on more different substances than other common bonding agents and is well suited for bonding glass, plastics, and metals. Straight epoxy is clear, but manufacturers add colored fillers; it cures at room temperatures at various rates depending on the formulation. Epoxy resin seems to resist moisture as well as any exterior woodworking glue.

14

Wood Finishing

Finishing in this book means *coating for protection and beautification.* Paints are coats that obscure wood, and *painting stands for covering with paint.* I consider painting a kind of finishing, though some authors restrict *finishing* to *clear* coatings through which wood remains visible. Generally, the coating is applied as a liquid that hardens (*dries*), but thin solid sheets (*overlays*) bonded to the wood surface serve essentially the same purpose and are here treated as a variety of finish.

14.1. WHY AND WHERE WOOD NEEDS FINISH

All finishes, clear coatings as well as paints, change the appearance of wood. The clear kinds modify light reflection, with the effect of deepening wood color and enhancing the figure of annual rings and other wood features. Proper finishing improves appearance and preserves it as well, since unfinished surfaces deteriorate—in different ways indoors and outdoors as time goes by.

Weathering

Under outside exposure many wood species first turn brown, due to washed up extractives, before the surface *weathers* by the action of light, moisture, and fungi. Direct and indirect sunlight causes photodegradation: ultraviolet rays slowly break lignin and other wood constituents down; some of the fragments oxidize, and fiber bonds become weak. Large moisture fluctuations cause surface layers to swell and shrink—in addition to contraction due to degradation—while stable deeper layers restrain these movements, so that weakened fiber bonds fail and the surface checks. After rain, and in damp weather, spores of fungi germinate on the surface; their hyphae grow among the loosened cellulose fibers, in checks and in wood cells, living on cell contents and degraded wood constituents. The fungi spread nonuniformly as moisture permits and give the surface a blotchy appearance.

Weather-exposed wood surfaces erode at an average rate of 1/4 in. in 100 years (Fig. 14.1)—porous earlywood faster than dense latewood, so that end grain and radial surfaces of many woods corrugate (Fig. 14.2). Weathering brightens woods of dark color and turns light-color species darker, but in the second year all species turn gray. Where no rain washes black spores of fungi and oxidized wood fragments out, the surface darkens, becoming deep brown or black. This is why on high mountains, where most precipitation falls as snow, wood structures look dark. Another reason is that in high altitudes the thin air filters out less destructive ultraviolet light.

Indoor Changes

Light degrades and discolors wood indoors also, but more slowly and in different ways. The light there is less intense, and glass retains substantial portions of ultraviolet. The effect appears most quickly on cheap paper, which yellows mainly from photodegradation of lignin, the second largest wood constituent. Cellulose, wood's major constituent and the sole constituent of good paper, is more stable. Common clear finish does not prevent photodegradation, so that practically all natural wood surfaces change slowly as time goes by. This poses problems when furniture and paneling have to be replaced or supplemented with matching pieces. Some finishing systems include ultraviolet-absorbers to retard the change in color.

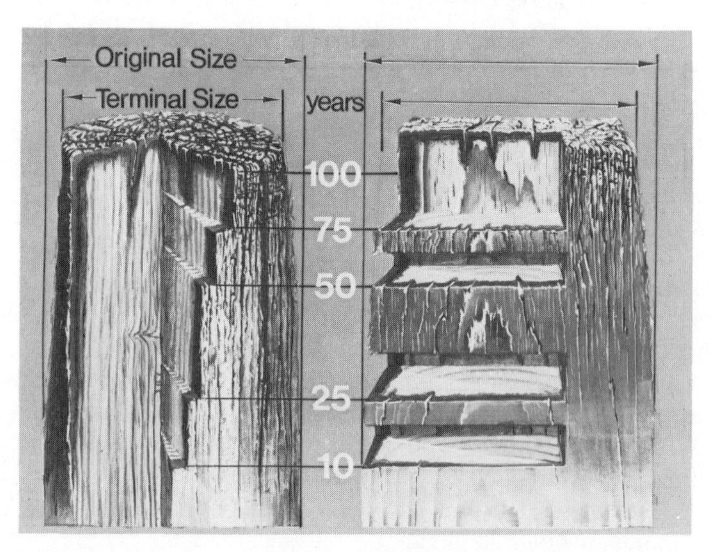

Figure 14.1. Weathering of wood. (Photograph courtesy of W. C. Feist and R. V. Paynter.)

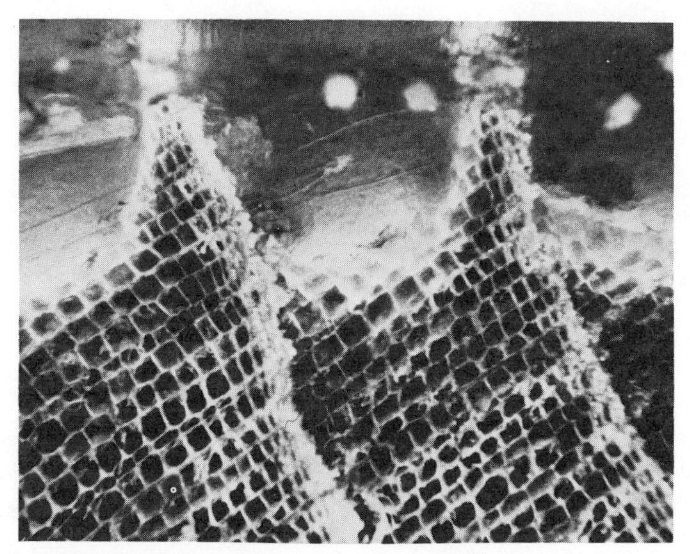

Figure 14.2. Microscopic view of a weathered western red cedar surface in cross section, showing corrugation. (Photograph courtesy of W. C. Feist.)

Interior wood has to be finished mainly because people touch wood, inevitably leaving fingerprints and causing spots from liquids, grease, and other pasty matter which penetrates pores and capillaries. Only sanding can clean such an unfinished surface. Finished surfaces are smooth and so *tight* that they stain very little and impurities are easily wiped off. The main purpose of interior finish is to provide a cleanable surface.

14.2. KINDS OF FINISH

Finishing systems are legion. Covering them all is not possible. Instead this section focuses on the cardinal features of the various finishing systems: transparency, film-forming versus penetrating, permeability, and smoothness.

Paint Versus Transparent Finish

Features in wood such as annual rings show by the same mechanism as other objects: they reflect light to the eye, and latewood reflects differently from earlywood. Paints let no light or only a little light through; the fraction which does pass and is reflected gets absorbed by the paint on its way out, so that the wood itself remains invisible.

What makes paint opaque is its pigment—mostly a powder of opaque minerals. Pigment particles are nearly impermeable not only for light but also for gases, so that water vapor on its way through a paint layer has to diffuse around the imbedded particles on a crooked, long, narrow path. Because of its pigment, paint seals basically better than clear finish. Aluminum paint contains flat aluminum flakes which overlap and make the paint especially tight.

Light absorption and low permeability contribute to the durability of finish. Paint therefore lasts longer than clear finish film, which may remain intact itself but may separate from the degraded wood base. Dark paints absorb light better than light paints and are somewhat more durable. Under outside exposure, the lifetime of good paint is about eight years, while clear film-forming finish fails in less than two years.

Stains

Stain stands inbetween paint and clear finish: it adds tone but leaves the wood figure visible, and may even enhance the figure by concentrating in a part of the annual ring and in rays.

How well stain protects depends on the amount of pigment. In most types pigment particles provide the color; in others stain components react with wood constituents and form colored compounds. The latter types have little effect on photodegradation and no effect on vapor passage, while in the common type the pigment particles form a thin line of resistance against light and diffusing vapor. Stain protects less than paint also because it is not essentially a film-forming finish.

Film-Forming Versus Penetrating Finish

All finishes coat the wood, and all to some extent fill voids of the microscopically rough wood surface (Fig. 14.3). Film-forming and penetrating finishes differ in thickness and in the shape of the coating. Finish films appear as a distinct layer on the surface, sufficiently thick to be visible with the unaided eye in a cross section, their surface very even and not following the rough wood contours. By contrast, coatings of the penetrating kind parallel the microscopically rough wood surface contour and are too thin to be evident. Films can be lifted off, penetrating finishes cannot. Some authors see penetrat-

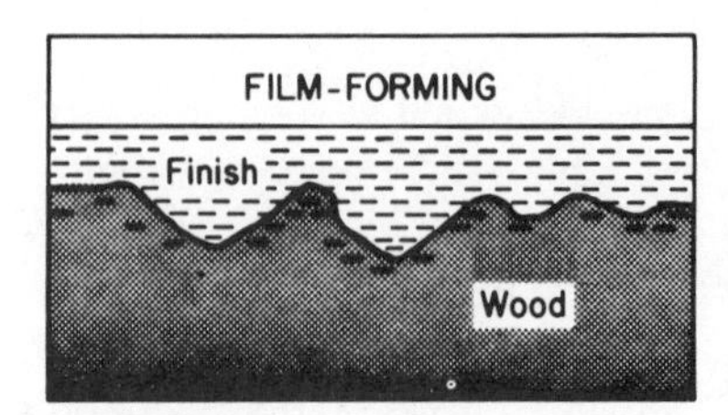

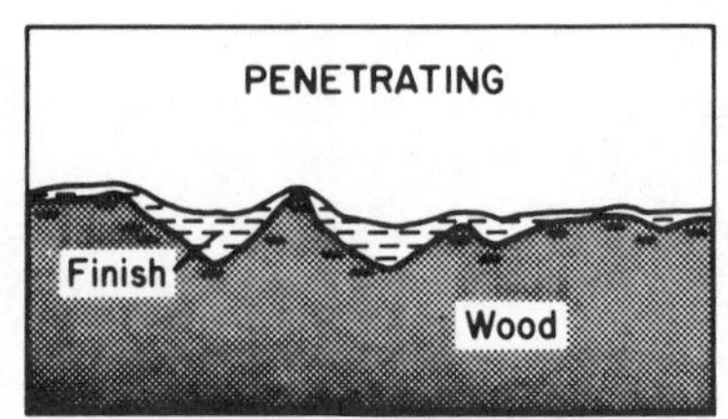

Figure 14.3. Schematic illustration of film-forming and of penetrating finishes.

ing finish as a noncontinuous coating, although liquid flowing down from the tiny surface ridges still leaves a thin coating on these high points.

Since thin coatings offer little resistance to the elements and to wear, films are more durable than penetrating finish. However, in some cases the penetrating kind may last longer. Wood and finish coatings tend to swell and shrink by different amounts in the directions of the surface plane. Wood forces its movements on the coating, causing shear stress at the interface of coating and wood. To stretch and to compress a film requires much more force than to stretch and compress invisibly thin coatings. The shear stress under the film is consequently much higher and more likely to tear the two materials apart. Films also peel and blister; penetrating finishes cannot. Both finishes crack, but the cracks remain invisible in the penetrating kind. In some cases, the latter retain good appearance longer than a film. One may succinctly say *no film—no failure.*

Permeability

Another question is how well and how long the finish protects the wood. Surface films are superior in this regard, but after cracking and peeling occurs, protection depends on finish within the wood. Then the penetrating kind, specifically formulated for protection within, may again be more effective.

Most coatings are impervious to water, which consists of clusters of molecules. By contrast, no finish is sufficiently tight to prevent passage of the individual water vapor molecules. The *Wood Handbook* gives numbers of *moisture-excluding effectiveness,* which were obtained by exposing finished dry wood samples to humid air. The numbers range from zero for one coat of water-repellent preservative (WRP) to 95% for a 3-coat aluminum paint, meaning that WRP had no effect and that the amount of vapor adsorbed by the painted sample was five percent of that adsorbed by an unfinished sample. The numbers have only representative value; under different conditions, particularly with samples of different size and species, the effects would vary. Under exposure to certain humidity for several months, all wood regardless of finish type approaches the equilibrium moisture content of the atmosphere.

Smoothness

Porosity and smoothness of the finish determine the degree of gloss or luster and cleanability of the surface. From the point of view of sanitation the finish cannot be too smooth and too close. However, uniform light reflection causes smooth surfaces to shine, giving distinct gloss or sheen, which is often perceived as showiness and as unpleasant glare. Smoothness has shortcomings.

A penetrating finish remains almost as rough as the wood surface; it appears mat and shines little more than unfinished wood. Finish films are relatively smooth by nature, the particular degree of smoothness depending, among other things, on pores that solvents leave behind when escaping from the drying film. Porous, *breathing* film looks mat.

Dry porous films can be smoothened mechanically by sanding, rubbing, and polishing operations which fill film pores and wear down peaks until the surface has the desired sheen. Inversely, rubbing with fine abrasive dust eliminates gloss by scratching numerous tiny valleys into the smooth surface.

Natural Finish

Clear finish is designated *natural* because it accentuates and preserves wood's original appearance, but strictly speaking the film-forming clear kind is not *natural* since it lacks natural roughness and luster. For outside woodwork, natural finish in the restricted sense of the penetrating kind is known as *natural exterior finish,* whereas for inside conditions more specific names like *Danish oil finish* (see Section 14.8) are in use. *Stain* qualifies as *natural* if it only intensifies the color of the particular wood species and does not add another color.

14.3. EFFECT OF WOOD PROPERTIES ON FINISH DURABILITY

Porosity and affinity to other substances make wood a good finish substrate. Each kind of wood and all wood-base panel products can be finished, though particular coatings and finish systems have to be selected for some. This section shows how different wood properties affect finishes and their lifetime.

Swelling and Shrinking

It has been explained in Section 14.2 how swelling of the substrate contributes to separation of finish films. Actually a small degree of swelling, such as in fiber direction of lumber, does not matter and may even prolong the film's service life, since films swell too. But large wood movements, especially high tangential swellings, strain and break the coating. Flat-grained lumber is inferior to edge-grained lumber as a finish substrate, because it swells most in direction of width (Fig. 14.4). Widthwise swelling and shrinking of siding boards cannot be restricted by nailing near both edges because the boards would break or buckle; use one nail per bearing at a spot where the nail permits board movement.

Flat-grained material is inferior also because dense latewood tends to swell more than porous earlywood. On edge-grained surfaces, latewood swelling may cause slight unevenness without consequences for the finish film, but on the pith side of flat-grained lumber swelling of the wide latewood shells leads to shear stress and separation of latewood from earlywood (Fig. 11.7). The finish film fails above the juncture.

Penetration of water into wood siding depends on the orientation of the

Figure 14.4. Photo of house siding showing how paint on edge-grained boards lasts longer than on flat-grained boards. (Photograph courtesy of W. C. Feist.)

wood fibers. The fine grooves formed by wood cells and cell cavities hold more water on horizontal boards than on vertical boards. The difference becomes striking on weathered wood when erosion of soft earlywood creates the *corduroy surface* shown in Fig. 14.2.

Wood surface checks tend to open and close as moisture content changes, causing cracks in the finish film. Flat-grained boards with their large width-directional movements, in addition to having weak planes of rays perpendicular to the face, check the most and are for this reason inferior again.

For reasons explained in Section 6.1, plywood checks even more than flat-grained lumber and is considered unpaintable if not overlaid. Plywood without overlay lends itself better to penetrating finish. If painting is nevertheless required, it should be done in the dry state with checks at their widest whereas for lumber a state of above-average moisture content is recommended during painting (Section 14.6).

Cracks in finish films from substrate defects rarely occur on particleboard and fiberboard; these two panel products represent a good film substrate, at least when pressed for dense surfaces that absorb little more finish than solid wood of the same density.

Structure-Related Properties

Flat-grained lumber with distinct annual rings has another disadvantage: finish films hold poorly on dense latewood because its small cell cavities provide little space for mechanical adhesion; such tissue is a poor anchorage.

On edge-grained material the film easily bridges the narrow latewood edges, but it cannot straddle the wider bands of flat-grained lumber.

Earlywood-latewood contrast obviously contributes to failures of finish films. Southern yellow pine with its dense and wide latewood bands is worst in this regard, while the diffuse-porous hardwoods fare best, and species like northern white pine with indistinct annual rings rank inbetween.

Generally finishes last the longest on low density species which swell and shrink little. In the case of penetrating finishes, they last longer also because light wood species absorb more of the liquid finish.

The rough, extremely porous end-grain surfaces are difficult to finish because the liquid sags into this substrate. For film-forming, the pores and depressions need pre-sealing with varnish, for example, before the actual finishing. Fortunately ends of lumber are rarely exposed and seldom need finishing; but round and oval knots at lateral surfaces have end grain too, and must be treated accordingly to prevent excessive finish absorption and achieve the desired gloss.

Large pores of species like oak and the lauans drain still-liquid finish out of surface coatings, causing air bubbles and pin holes which contribute to film failures and distract from appearance. In interior finishing of better quality large pores have to be plugged with powder, paste, or a liquid *filler.* In products such as guitars, whose value depends heavily on appearance, even the relatively small pores of birch, for instance, require filling. The small cell cavities of softwoods and of many fine diffuse-porous hardwoods need no filling; these species are for this reason easier to finish than ring-porous hardwoods.

Chemical Aspects

Extraneous wood constituents such as resin of pines and oil in teak interfere with many finishes, especially above the end grain of knots when heat forces these constituents to the surface. Resins and oils may retard finish drying, cause tackiness, and lead to unsightly blotches. Surfaces of such woods require cleansing with solvent before finishing. Commercially available *knot sealers* minimize the *bleeding* by obstructing the pores in the surface. High drying temperatures force free resin and oil out and *set the pitch,* which is then planed off before finishing.

For house siding and outdoor furniture, redwood and western red cedar offer two excellent finishing characteristics: attractive brown-red color and shrinkage that is low even for the low density of these species. Both color and dimensional stability result from large amounts of heartwood extractives, which have however a severe disadvantage in causing paint discolorations (Section 14.4 and 14.5).

Wood preservatives may interfere with finishing when the finish dissolves the preservative or when the preservative solvent had not evaporated before finishing. Creosote and other preservatives that remain liquid tend to stain through paint if they have not weathered out before painting.

14.4. PAINTS AND PAINTING

Painting (and particularly painting of house siding) deserves special attention because many people paint and repaint the wood of their home themselves. Besides which, paint and painting are well suited to illustrate concepts of other kinds of finishes and finishing.

Composition of Paint

Liquid paint consists of solid *pigment* particles, which impart the color, and a liquid portion, the *vehicle.* The latter is made up of the *binder* and, though not always, a volatile *thinner* or *solvent* to facilitate spreading. Aluminum paint derives its name from the pigment, while other paints are named after the vehicle—or after the physical state of the vehicle—as being the component that determines all essential paint properties except color. The vehicle mainly converts the liquid into a solid; its nonvolatile binder represents the essential film-forming component which holds the pigment particles together.

Since the pigment is less permeable than the binder, and paint has less permeability than clear finish film, the reader may conclude that increasing amounts of pigment make paint tighter. This conclusion is correct only within limits. Paint with much pigment has not enough resin to bond the particles; it is porous, of the *low luster* or *breather* type. A large percentage of resin makes paint tight and gives a continuous, smooth, glossy surface. The resin percentage should be at least one half of and at most equal to the percentage of the pigment.

Oil-Base Paint

Traditional oil-base paint contains linseed oil or other organic oils as binder. Most modern paints of this kind are modified with alkyd, a synthetic resin. The strong smell of oil-base paint results from evaporating thinner, typically petroleum spirits or turpentine.

The evaporation thickens the remaining liquid but the paint dries mainly by oxidation of the air-exposed binder. Small amounts of factory-added *dryers,* usually oil-soluble metallic compositions, accelerate the oxidation process. Nevertheless, the paint dries very slowly because it takes hours for oxygen to diffuse into the coating. Oil and oxygen also have no great affinity for each other. Slow drying is one reason why oil-base paint penetrates well; therefore it is a good *primer,* meaning, well suited as a first coat. Primers should be tight, oil-base paint can meet this requirement.

Latex Paint

Latex paint gets its name from being an aqueous emulsion like latex and looking like latex, the milky juice given off by certain plants. (Rubber latex from rubber trees serves as raw material for tires and other rubber products.) Actually *stable emulsion* or *stable dispersion* would be more descriptive than *latex, stable* implying that components do not separate during storage (although latex paint still requires shaking because the pigment will separate). Besides water, the main solvent, most latex paints contain small amounts of other solvents.

Pigment-free latex finish resembles white glue and becomes clear while drying. The binder is either acrylic resin or the less expensive polyvinylacetate (*vinyl*) resin. In the liquid paint tiny droplets of resin and particles of pigment float in water. When water evaporates out of the spread paint, resin droplets and pigment particles float together and bond on touch. The paint quickly solidifies without chemical reaction; its resin droplets have too little time and are too large to penetrate fine wood capillaries, only the solvent water does that. Already for this reason latex is not a good primer. To prevent overly rapid migration of water out of latex paint into the substrate on warm days, some painters hose the wood surface down shortly before painting.

Dry latex paint retains such large, formerly waterfilled pores that it cannot be used as primer on redwood and western red cedar. Water-dissolved wood extractives migrate through pores to the surface and cause discoloration, as explained in the following section. The paint has another disadvantage: below 50°F the resin droplets are too stiff to bond properly.

However, latex offers two advantages that outweigh the shortcomings: it excells in durability and ease of application. Ultraviolet light destroys latex paint at a slower rate than oil-base paint. Of course, this paint is not suited for all purposes and not each latex paint is good. Since the solvent water costs nothing, the paint can be made cheaply with a high proportion of water; it should contain no more than 40 to 50% water.

Painting of House Siding

Painting procedures vary with the substrate and with the kind of paint. However, for painting new weather-exposed wood three steps can be recommended.

Step 1: Apply Water-Repellent Preservative First brush on a water-repellent preservative (WRP, described in Section 14.7) or apply it with an oil squirt can. The repellent is needed especially where, under defective paint, the wood later would soak up water: at end-grain faces and at all joints, particularly those of end grain surfaces. Water drips have to be treated too, so one may as well brush the whole surface. Water-exposed millwork receives the treatment before assembly in the factory. Window corner joints and the groove at the edge of the glass pane both outside and inside, particulary along the lower rail, definitely need water-repellency.

On bare wood the repellent does not interfere with following paint coats since it contains only a small percentage of wax and it penetrates well. In repainting, however, any WRP left on remaining old paint surfaces should be wiped off. Repellents protect the wood for several months before priming and final painting.

Step 2: Priming Priming and caulking may follow as soon as the WRP solvent has evaporated, after at least two days of favorable drying weather. The primer serves as the medium between wood and paint film. It should be flexible like a mattress and should equalize stresses in areas of stress concentration, such as at the edge of wide latewood bands on flat-grained lumber. Generally the primer has the color of the final coat but a different composition. Wood that holds paint poorly calls for a low-permeability aluminum primer. In some systems the same paint, called *selfprimer,* serves as primer and as final coat. In any case, brush on the primer sufficiently thick to obscure the wood grain.

Holes of sunken nails and joint gaps too large to be bridged by the primer need to be caulked. Since the pasty caulking compound does not penetrate wood, priming precedes caulking.

Step 3: Final Painting The final coat leaves no trace of the wood grain visible. In protected areas, such as siding under overhanging roofs, one coat—not counting priming—is enough, while exposed surfaces require two coats to bar water, and to have in reserve for wear. The film should be 4 to 5 mil thick, corresponding to one gallon of average paint (including primer) per 600 ft^2. Thicker film tends to crack (see Section 14.6). It is therefore wrong to repaint before the film wears down to less than 4 mil, but it is also wrong to wait until the wood surface appears and requires priming again.

All coatings can be applied by brushing, rolling, and spraying. Brushed paint adheres best, since strokes in fiber direction force paint into the grooves. Spraying is least satisfactory in quality and durability because it deposits little solids and lacks uniformity.

14.5. EARLY FINISH DEFECTS

Some defects result from aging of the finish and will be treated in the next section. Others, discussed here, have a variety of causes; they can develop at any time, but their early appearance soon after finishing is likely since the finish is disposed to them almost from the very beginning.

Discolorations

On redwood and western red cedar siding some paints discolor when water carries brown heartwood substances to the surface. Such extractive staining occurs in two different ways. In the first, water penetrates through capillaries of porous paint, soaks into wood, dissolves extractives, and later during drying carries solutes to the surface. Paint *bleeds* this way rather uniformly in a diffused pattern, mainly where rain and dew wetted the siding. Thin paint and lines of thin paint on tiny ridges of rough wood surfaces tend to bleed most. Paint that bleeds is obviously porous, especially in the primer, and lacks a water-repellent seal underneath.

Run down discolorations result from water that enters the back side of siding boards, either from roof leaks and behind ice dams, or from cold-weather condensation in walls without a vapor barrier. The water seeps down behind and within the boards carrying extractives along; it runs out through joints and washes the solutes to the surface, where they appear as characteristic brown streaks.

Other discolorations develop during long periods of warm humid weather from mildew or mold fungi (Section 10.2) that grow on the finish. Fungicides prevent this defect, but mercurial compounds, the most effective fungicidal finish additive, are not allowed anymore. After mold occurs, kill the fungus with household bleach and restore surface color at the same time. Conditions which favor mold may also cause blue stain in wood (Section 10.2); paint conceals the stain, but if wood stays damp for a long time the fungi push through paint to the surface.

Blisters

Paint blisters (Fig. 14.5) are explained either by pressure of gas and liquid at the interface of wood and paint, or by excessive swelling of flexible fresh paint in the direction of the surface plane.

Figure 14.5. Paint blisters on house siding. (Photograph courtesy of W. C. Feist.)

Temperature blisters result from expansion either of paint solvent or of air in wood that was painted with dark paint at low temperatures, before the sun had heated the surface. The blisters appear several hours after applying a thick final coat whose skin hardened while the deeper part was still soft. Temperature blisters on siding can be avoided by painting after the sun has heated the surface, or while wood cools. Start painting the north side in the morning, and follow the sun around the house. Painting late into evenings involves another risk when, during cool nights, dew from humid air swells the paint, causing wrinkles, fading, and loss of glossiness.

Moisture blisters result from wood that was too wet at time of painting and did not provide adequate adhesion. Pressure in the blisters may come from water which evaporated under the warm, new, still flexible film of paint. Moisture blisters were quite common when high-swelling primers with zinc oxide pigment dominated the market.

Raised Grain

Grain-raising means either *raising of fibers* or *of annual ring shells.* In the sense of *raised fiber* it is a wood surface roughening from the gain and loss of wood moisture, when loose fiber ends and fiber fragments—attached to the surface or pressed into pores—rise and stand out from the surface. Wood that was planed or sanded in the rough wet state has a strong tendency to grain-raising of this kind. The raised fibers protrude through the liquid finish film and appear immediately after finishing. The raising must be expected whenever recently planed or sanded surfaces swell under the influence of the solvent water of latex and aqueous stains. *Solvent* finish rarely raises the grain.

Raised grain in the sense of *raised fibers* can be avoided by wetting, redrying, and sanding slightly before applying finish. Outdoors, the wetting procedure may also be practical because wood moisture content at the time of film application should be above average, as explained in the following section under *fissures.*

On the pith side of flat-grained lumber, dull planer knives press dense latewood bands down into the porous earlywood layer, whose soft cells yield like a cushion (Fig. 11.6). Later when swelling and shrinking, the compressed earlywood cells recover and push the latewood band up, leading to stress and cracking in the finish film at the edge of the latewood band. Such grain raising in the sense of *raised annual rings* appears mostly on old finish films.

Peeling

Peeling is a separation parallel to the film surface, either in the wood interface (*interface peeling*), or within the paint film (*intercoat peeling*). The first kind is attributed to moisture: paint cannot hold on wet wood with water-filled voids; at time of painting the wood moisture content should be below 25%. Paint also peels when wood absorbs water long after painting—

just how, is not well understood. I explain it, figuratively, that wood moisture migrates to the surface and pushes the film off. The water may enter through paint cracks, in other cases it seeps in through porous paint, or it comes from behind. Siding boards soak up water at the back where it runs down from the edge of the roof, or rises from wet ground by capillary action.

The cause of peeling at the wood interface is less obvious if water has condensed in the walls or under the roof, from there trickling down into the wall. Peeling from water condensation is now much more common than it was before people insulated their houses and kept them tighter to save fuel. Vapor barriers as described in Section 3.10 prevent condensation in walls. Installed in ceilings they would prevent roof condensation the same way, but it is more practical to ventilate the roof space, a measure effective also against ice dams, another source of water in walls. On unventilated roofs, heat from inadequately insulated ceilings melts snow, water flows down to the cold overhanging roof ends where it freezes. This resulting ice dam backs up further melt-water, causing water to seep through shingle joints into the wall. When snow has melted on southern roofs, the sun then heats the whole attic and contributes to an ice dam on the north side, where the problem is from the beginning more severe.

As will be explained in the following section under *chalking,* at the surface of aging paint a soapy layer develops on which new paint, especially latex, does not hold well, causing intercoat peeling. Therefore paints should not be more than two weeks old before applying the next coat. Older coats have to be brushed and cleansed with detergent solution. If in doubt, check the bond by pressing self-sticking tape (such as band aid) on a test spot of second-coat paint one day old, and quickly snap the tape off: it must come away free of paint, the new coat must stick to the old coat and not to the tape.

14.6. AGING, FISSURES, AND CHALKING OF OIL-BASE PAINT

Physical and chemical changes in finish do not end with drying; they continue indefinitely though at slow rates and are the main reason why outside finishes do not last. Under exposure to the elements all finishes are prone to deterioration. Aging and related defects have been thoroughly researched for oil-base paint; other finish films undergo similar changes.

Finish-Film Aging in General

Oil-base finish dries by *polymeric oxidation,* meaning *oxygen from the air combines with the paint vehicle to form polymers.* After drying follows *destructive oxidation,* a process that embrittles and stiffens films by destroying plasticizer, the substance that renders the finish film soft and flexible. Embrittlement results also from *cross-linking,* that is, chemical reactions between parallel molecular chains during formation of larger complex molecules.

Some of the oxidation products are volatile and escape, causing the remaining components to pull closer together. Crosslinking ties them further; the film contracts and becomes more dense. In the surface plane the dry wood substrate restricts this contraction, so that the film stretches in this plane and contracts in thickness only. In fact, restraint in the surface plane causes increased thickness contraction, just like a rubber band becoming narrower when stretched.

Sunlight accelerates finish aging; but actually aging in darkness and *photodegradation* in light cannot be separated. Moisture greatly speeds the process, especially of physical changes. Finish has affinity for water; some of the oxidation products are distinctly hydrophilic. Finish adsorbs water and holds moisture in varying amounts depending on air humidity and moisture content of the wood substrate. Rain water and dew naturally raise finish moisture significantly. As in wood, moisture adsorption and drying of the finish involve swelling and shrinkage, which the wood again restrains in the surface plane, causing additional stresses as well as permanent stretching and compression [like the plastic stretching and compressing of wood layers in casehardened lumber (Section 3.9)], with further embrittlement as a result.

Fissures

Under stress, embrittled and rigid paint finally reaches the fissure stage, checking and cracking (Fig. 14.6). Checks are superficial fissures, while cracks reach down to the wood. The cracks either extend at once throughout the depth of the film, or they start as checks which gradually deepen to the wood surface. Both failures occur most commonly when the finish film is dry and the wood swollen; they run predominantly in fiber direction, perperpendicular to the direction of wood swelling, and parallel to pigment particles as they were oriented by the brush strokes. Thick finish layers tend to crosscrack in conspicuous, coarse, reticulate fissures.

The finish film does not crack if the wood contracts with the film. To this end one should apply the finish at a moisture content that is above the expected average. For house siding 15% appears to be optimal in most parts of North America. When after finishing the wood dries and shrinks more and faster than the film contracts, the film comes under compression but does not fail, and later can expand without coming under tension stress.

Siding boards painted at moisture contents far above 15% tend to check when their surface layer dries and shrinks rapidly. The paint cannot be expected to bridge the gap since a finish sufficiently flexible for that would be too soft. Consequently film cracks that extend into wood can be assumed to result from checks in wood; they expose deep wood layers so that under wet conditions the finish rapidly deteriorates.

Figure 14.6. Fissured paint on house siding. (Photograph courtesy of W. C. Feist.)

Paint Chalking

Paints age in the binder, not in the pigment. The superficial layer of binder erodes, laying bare a granular bed of pigment, so that the surface becomes rough. Paint contraction accentuates this effect. The rough surface reflects light in a diffused manner and the paint loses gloss. Penetration of light into surface voids between pigment particles causes diffused reflection and affects the paint's color, making it less distinct and impairing its brilliance—the color fades.

Deprived of binder the pigment particles loosen and rub off, the paint *chalks* and wears down. Some binder decomposition products are fatty acids, a sort of soap which if not washed out by rain, accumulates on the surface as a soapy layer and causes the above-mentioned intercoat peeling.

Identifying Finish Film Defects on Siding

Since moisture and light contribute a great deal to finish defects and differ on house siding in quantity and intensity, respectively, on the various sides, a little detective work can uncover their cause. Observe where and when the defect developed. Paint peeling in the early spring outside bathrooms and kitchens, for example, indicates that moisture from cold-weather condensation inside the wall caused the failure.

14.7. CLEAR FILM-FORMING AND PENETRATING EXTERIOR FINISHES

Of these two, penetrating finish is by far most important, since the clear film-forming type does not last under outside exposure. For penetrating finish any surface is suited; rough sawn and weathered surfaces because they absorb more finish may be even better protected than new planed surfaces. Penetrating finish is easily renewed since there is no old layer and no chalk to be removed; only discolorations from extractives and fungi may have to be washed off with a detergent. House siding finished with the penetrating type needs refinishing when it starts to look blotchy from uneven growth of fungi, or when the natural weathering process begins to show in more exposed parts. This section covers first the main systems of penetrating finish, before dealing with the clear film-forming kind.

Water-Repellent Preservatives

Water-repellent preservative (WRP) as a very thin, not film-forming liquid is of all finishes easiest to apply. It contains a fungicide (usually pentachlorophenol), some resin, and a small amount of wax—all dissolved in mineral spirits that represent the bulk of the mixture. You can prepare WRP at home, but the mixture is more readily available and hardly more expensive than its ingredients. Most lumber dealers and paint stores carry the popular brand *Woodlife.*

WRP has a *large spreading rate,* meaning *a unit volume covers a large area.* Small pieces can be dipped to treat all surfaces. On siding, all-surface treatment prevents water from penetrating through lap joints and crawling up at back surfaces by capillary action. When brushing installed horizontal lap siding, spread the repellent liberally along the lower edge to let the finish work up into the lap between boards.

The solvent evaporates and leaves resin, wax, and the preservative in the surface. The wax blocks capillary flow, repells water (Fig. 14.7) and protects very effectively during short periods of wetting; for aboveground purposes it is the most important WRP ingredient and protects without preservative. On new-planed and sanded wood, WRP's last one to two years; when renewed, the weathered surface absorbs more than when unweathered and remains protected for about two years.

Some penetrating finish contains relatively large amounts of wood preservative and sells as *fungicidal finish.* This kind is supposed primarily to prevent discoloration from growth of fungi in the wood surface. Manufacturers may or may not add wax, and it depends on this additive whether fungicidal finish falls into the category of WRP's.

Stain Systems and Staining

Some finish stains are oil-resin-based, others represent a latex system, and a third kind resembles WRP; all contain either some pigment or an ingredient which reacts with wood to form a colored compound. The pigment absorbs light and protects by preventing photodegradation of wood constituents and of the preservative. For this reason, and due to a higher percentage of wax, stains last longer than clear penetrating finishes: eight to ten years on rough,

Figure 14.7. Effect of water-repellent. (Left) Treated wood surface. (Right) Untreated wood surface. (From Feist and Mraz, 1978.)

weathered surfaces. On the other hand, the relatively large amount of wax interferes with adhesion and drying of later coats, should the homeowner want to paint instead. The wax may even interfere with renewed stain coats.

Staining is not quite as easy as brushing on WRP, because on large surfaces stains may cause *lap marks* when staining is resumed at an edge where stain has dried. The overlap shows up in more intense color, since the dry first coat seals and prevents penetration of the second coat. Work on wet edges before the front edge of the stained area dries. On walls consisting of sections the mistake can be avoided by stopping at section ends only. Where a second coat is required, apply it before the first dries. Excess stain at the wood surface shows as a glossy spot; wipe it off before it dries.

The well-known *Forest Products Laboratory Natural Finish* is a stain with pigment in the color of the particular wood species; hence the designation *natural. Bleaching stain* contains a gray pigment for the appearance of weathered wood and also chemicals which bleach wood like weather does.

True *latex stain* is on the market, but the industry also uses the name *stain* (as well as *heavy-bodied stain, solid color stain,* and *opaque stain*) for latex finishes which are not stains: they form films and are opaque or nearly opaque—two characteristics of *paint* but not of *stain.* Manufacturers and dealers misuse the magic word *stain* as a sort of sales gimmick.

Protective Salts

Inorganic salts have been used in stains for many years to generate color by reactions with wood constituents. Other inorganic salts served as waterborne wood preservatives, although for the fungicidal component in WRPs, manufacturers prefer pentachlorophenol as long as the Environmental Protection Agency will allow it. More recently researchers tested numerous inorganic salts as main finish components; what they found is encouraging.

The salts are fungicidal, preventing growth of mold and of other fungi with their resulting discolorations. Some have no color, but the most interesting types, for instance certain copper chromates, impart a light green-brown to the wood. The copper chromates and several other explored salts absorb ultraviolet light with the effect of protecting wood from photodegradation.

The salts are applied in aqueous solution. After soaking and diffusing into wood, they react with the substrate and become insoluble; neither rain nor groundwater leach them out. They also seal off moisture and reduce swelling, thus retarding weathering. In addition, salts prevent the discolorations of redwood and western red cedar by fixing the extractives.

With all these favorable characteristics the inorganic salts represent a complete finish system in their own right and give good service without additional components. The kinds that tone the wood while keeping the figure visible may be considered as a *stain*.

The new chemicals have beneficial effects also as a minor part in other finish systems. Researchers found out that these salts brushed on wood before painting contribute to durability of paint, though the mechanism of this effect is not fully understood as yet.

Clear Film-Forming Exterior Finish

Traditional transparent finish films last outdoors less than two years. Their durability is limited in the first place because light passing through them degrades the wood surface and dislodges the film. The finish substance itself degrades too, but at slower rates. In recent years two breakthroughs brought durable transparent finish films within reach: the protective salts discussed above and new polymer film coatings that resist photodegradation.

The salts are brushed on in aqueous solution and allowed to dry before spreading the polymer. The latter, silicone resin, for example, does not absorb ultraviolet light, rather lets the light pass through to the salt-protected wood and so remains intact. Finish systems of this type may last five years or more, but are not on the market, at the time of this writing.

Wood siding under any kind of natural finish browns and blackens around nails, from which unsightly streaks extend over several inches in fiber direction. The cause of this nail staining is obvious: in acid-moist wood, common nails inevitably corrode, the rust dissolves and diffuses in water of the surrounding tissue where it reacts with wood constituents, forming dark compounds—reacting especially with the tannin of oak to iron tannate, a substance used in making black ink.

Under penetrating finish, nail discolorations show up as early as the first prolonged wetting, when the thin coating breaks at the nail. Films protect a little longer. Aluminum and stainless steel nails are safe, while galvanized nails corrode too, though more slowly than the common type.

14.8. INTERIOR FINISH

Since inside buildings wood moisture contents fluctuate little and the glass-filtered, relatively weak light is low in ultraviolet, interior finishes last practically forever; however, they must meet high standards in appearance and cleanability, especially on furniture. People want perfect surfaces, matching in color, very smooth, with particular degrees of gloss—requirements that render interior finishing difficult. Anybody can paint house siding, but few have the knowledge, skill, and patience to finish furniture.

The difference between interior and exterior finishing begins with surface preparation: while outside almost any clean surface is good enough, interior wood must be sanded with fine sandpaper, always along the grain to avoid transverse scratches. The circular sanders attached to power drills can therefore not be used. Further finishing steps depend a great deal on the particular type of finish.

Lacquers, Varnishes, and Varnishing

Lacquers typically contain nitrocellulose binder dissolved in an organic liquid, which evaporates after application and causes the drying. The solvents evaporate quickly and leave little time for penetration: lacquers are supposed to form films, difficult to brush, usually sprayed, and preferred in industrial production. We know them in the clear form and with pigment similar to paint.

Varnish forms a film like lacquer, but is clear by definition. In a narrow sense *varnish* means *oil* or *oleoresinous varnish* containing resin and drying oil dissolved in turpentine or in petroleum products, and drying primarily by chemical reaction. *Spar varnish* is an oil varnish for exterior use whose name originates from its use on spars of ships. *Spirit varnish* dries primarily by solvent evaporation and resembles lacquer, but its basic ingredient is not nitrocellulose. *Shellac varnish* represents a kind of spirit varnish, consisting of alcohol-dissolved lac, a resinous material secreted by an insect that lives on the sap of certain trees in southern Asia.

Varnishing requires the most care of all finishing operations. Preparation of the surface is quite demanding, as the wood remains highly visible under this blemish-accentuating coating. Besides fine sanding it may be necessary to stain and bleach for certain color effects before filling the pores. The varnish is carefully brushed or rubbed on in several uniform coatings, free of dust and air bubbles. Each dried coating has to be slightly sanded, in some cases with soapy water as a lubricant to provide a smooth *close* base for the following coating.

The last coating should dry about one week before a final rubbing eliminates dust specks and produces the desired degree of gloss. Cabinet makers rub with a mixture of fine pumice stone powder and non-drying oil on a cloth pad, followed by a second rubbing—always in fiber direction—with a still finer abrasive, and finally polishing with clean cloth. How much luster the finished surface has—if it is gloss, satin finish, or *flat*—depends on the rubbing and polishing, as well as on the kind of varnish.

Enamels and Enameling

Interior wood work such as cabinets could be painted, except the finish should be harder, smoother, and more glossy than paint: it should be enamel. This kind of finish is best known on the sheet metal surfaces of refrigerators, washers, and dryers, but protects and beautifies wood as well. Enamels differ from varnishes essentially only in the added pigment; consequently enameling by hand resembles varnishing in all steps described above. The various tight coatings and the pigment make enamel very hard and so impervious that even mildew cannot grow on it.

Penetrating Oil

Finish of high quality does no have to form a film. Some penetrating kinds—well-known from Scandinavian furniture—look as good or even better than films, the most popular kind being penetrating oil, which typically consists of drying linseed oil with little or no thinner. Application is simple: spread the oil liberally over the surface, allow to soak for half an hour or so, then rub in more oil with a pad of folded cloth, wipe off the rest, let dry over night, and repeat the treatment at least twice. This gives wood a very natural appearance with the mellow lustre of the porous wood surface, modified by the oil. This finish protects wood from staining and withstands hot dishes. *Danish oil* belongs to this category.

Floor Finish

Floors endure much more wear than other interior surfaces and deserve special attention. Floor finish can be of the clear film, paint, or penetrating type. This section focuses on the popular clear film wood floor, which costs more than most carpets, but lasts forever and so comes at lower cost in the long run.

Walking abrades floors by pressing grains of sand in the surface; the effect resembles sanding and wears the surface down. Actually wood wear in itself matters little. Millions of people can walk over unprotected dense hardwood before it wears out, as stairs and hallways in old public buildings show. Pressed-in dirt and tiny grains of sand change the color however from pleasant natural yellow or brown to dirty gray. To prevent this change, wood flooring must be coated with a finish film, which is flexible and high in resistance to abrasion. Among the many floor finishes on the market the more durable ones contain polyurethane resin as their main ingredient.

A thin coating of floor wax prolongs the life of the finish film. On a slippery waxed surface shoes find no grip and slide without exerting significant horizontal force; wax has filled surface voids with which peaks of the shoe-sole surface would otherwise engage and interlock. By the same mechanism the wax prevents, or at least reduces, scratch marks of rubber from soft shoe soles. Wax protects also by repelling water.

When finishing a new wood floor, first *seal* with a clear primer or *sealer,* a sort of thinned varnish which penetrates and provides a uniform base, preventing excessive absorption of further coatings. Then brush on two coats of clear floor finish, and finally wax after the last coat has thoroughly dried.

The finished floor requires little care. Being smooth and having no pockets to trap dirt, it vacuums easily. A damp cloth removes spots and persistent dirt. Scrubbing with detergent and rewaxing is rarely necessary. Under such care, floors stay beautiful for decades and have to be refinished only when occupants fail to maintain the shield of wax. If in more severe cases of neglect the film has worn down and the wood has turned gray, sand off the surface layer and refinish to achieve a new floor.

Wood-polymer flooring consists of wood whose cell cavities are filled with a polymer that is relatively hard and impervious. Some polymer penetrates the cell wall also and stabilizes the wood dimensionally to some extent. To achieve deep penetration, the industry presses liquid polymer or polymer precursors into the wood and triggers hardening by radiation. As no solvent has to evaporate, the polymer can be very compact, hard, and abrasive-resistant. Reports about performance of wood-polymer flooring vary. Some claim that it needs no protective finish film and becomes beautiful by the polishing effect of use. According to others the polymer as a sort of penetrating finish does not really prevent the change in color from pressed-in dirt.

14.9. PRINTED WOOD GRAIN AND OVERLAYS

Printed wood grain is a wood pattern (*figure*) printed on the surface of a less attractive material. The term *overlay* means a thin solid sheet, other than wood veneer, bonded to the surface of wood and wood-base panels. Prints and overlays belong together because both are applied predominantly by the industry and because wood grain is frequently printed into an overlay. As a rule, overlays and prints require no further finishing.

Direct Print Grain

This kind, printed on the substrate itself and not into an overlay, is commonly known in hardboard paneling, particleboard, and hardboard furniture (including cheap TV cabinets). If done well, the prints look like clear-finished genuine wood. The industry prints by means of engraved cylinders, whose pattern originates from photographs of real wood. The cylinders roll the printing ink down in lengthwise direction of the panels, so that wood pattern repetition appears in distances of about 5 ft, the circumference of the cylinder.

Some prints are imparted from engraved, plane press cauls that are free of pattern repetitions but can be recognized by the unnaturally uniform color of all prints. In large paneled rooms, two or more panels may have the same figure, but otherwise duplicates are rarely in evidence.

Printing requires a smooth uniform surface. Hardboard and medium-density fiberboard are well suited for this process, but still need a uniform opaque base or prime coat. Particleboard with flakes in the surface must be double base-coated. Species of plywood that lack *figure* can be printed without a base coat, but may need staining and filling since open pores and and rays show. A final clear finish protects the color, seals the print, and takes wear during handling as well as later in use.

Among the materials used in the various stages of grain printing the most noteworthy ones are polyester, alkyd resin, lacquer, and vinyl. The prints resist photodegradation better than most natural woods and change less under the influence of light.

Overlays

Like other finishes, overlays have several functions: some mask or hide defects, others beautify and decorate, and one group provides structural features such as hardness, check resistance, and stiffness in bending. All protect the substrate and give a tight, smooth surface. Good brands have sufficient strength, stiffness, and moisture-exclusion to prevent checking of the substrate.

Overlays consist of plastics, paper, or laminates as discussed below. Popular types are marketed as *contact-films* with a thin coating of a pasty binder under a peel-off sheet, while the wood industry applies the binder separately. Some brands have self-adhesive ability and bond when hot-pressed by flowing into the surface voids of the substrate.

Plastic overlays consist of polyvinylchloride or the more expensive and more durable acrylic polymers, produced either clear or in various colors by incorporating pigment. Up to this point, films of this kind seem to be essentially identical with liquid-system finish films, but being formed by melting, without solvents, they are less porous, harder, stronger, and more serviceable. Many plastic overlays carry wood grain in the color of the wood species made from blown-up photographs of real wood. The better ones feature embossed *wood pores* to simulate natural roughness and to create the genuine deep appearance of wood. This type is common on cheap furniture and TV cabinets.

Paper overlays consist essentially of strong paper, in the most simple form of one single sheet impregnated with phenolic or melamine resin for strength in the wet state. Many extremely thin veneers marketed for do-it-yourselfers are overlaid with a clear sheet of this kind, to give strength perpendicular to the fiber direction for shipment and handling. Veneer with overlays at both sides represents a composite sheet which shrinks and swells very little in direction of width, though more than plywood does.

Paper overlays with high resin contents are dense, hard, abrasive resistant, and clear or semi-opaque; marketed as *high-density overlays* (HDO) they need no finish. Sheets with less resin, the *medium-density overlays* (MDO) are opaque and intended as paint base. Plywood with MDO and HDO is produced for concrete forms, signs, and billboards. MDOs make plywood siding paintable.

The best overlays consist of several sheets, molded together to a dense infusible *laminate,* and are known under brand names like *Formica.* Within this category the decorative paper laminates with imprinted wood grain are most common. Each of their sheets has a particular function: the one nearest to the wood balances, some others provide a core, one seals out vapor, a decorative printed sheet determines the appearance, and finally a clear surface protects the print. Total thickness is typically 1/16 in. Surfaces of desks, good cabinets, and skins of heavy-use doors are typically of this kind. Laminates on kitchen and bathroom counters include a marmoreal printed sheet in place of the wood grain print. Where paper base laminates are not sufficiently flexible and would break when bent around corners, fabric laminates from resin-impregnated fabric layers may be worth the higher price.

Bibliography

American Institute of Timber Construction, *Timber Construction Manual,* 2nd ed., Wiley, New York, 1974. 778 pp.

American Society of Civil Engineers, *Wood Structures—a Design Guide and Commentary,* ASCE Structural Div., 345 E 47th St., New York, 1975. 416 pp.

American Society of Testing and Materials, *1979 Annual Book of ASTM Standards,* Part 22, *Wood, Adhesives,* Philadelphia, 1979. 1116 pp.

Anderson, L. O., *Condensation Problems—Their Prevention and Solution,* FPL 132, U.S. Forest Products Laboratory, Madison, Wisc., 1972. 37 pp.

Anderson, L. O., *Wood-Frame House Construction,* U.S.D.A. Agr. Handb. 73, Government Printing Office, Washington, D.C., 1970. 223 pp.

Baechler, R. H., *How to Treat Fence Posts by Double Diffusion,* FPL-013, U. S. Forest Products Laboratory, Madison, Wisc., 1963. 7 pp.

Beall, F. C., and H. W. Eickner, *Thermal Degradation of Wood Components—a Review of Literature,* FPL 130, U. S. Forest Products Laboratory, Madison, Wisc., 1970. 26 pp.

Browne, F. L., *Theories of the Combustion of Wood and Its Control,* No. 2136, U. S. Forest Products Laboratory, Madison, Wisc., 1963. 69 pp.

Browne, F. L., Understanding the Mechanics of Deterioration of House Paint, *Forest Prod. J.* **9** (11), 417-427 (1959).

Brown, F. L., *Wood Properties and Paint Durability,* U.S.D.A. Misc. Publ. 629, Government Printing Office, Washington, D. C., 1962. 10 pp.

Browning, B. L., *The Chemistry of Wood,* Robert E. Krieger, Huntington, N. Y., 1975. 689 pp.

Core, H. A., W. A. Côté, and A. C. Day, *Wood Structure and Identification,* 2nd ed., Syracuse University Press, Syracuse, N. Y., 1979. 168 pp.

Davidson, R. W., *Plasticizing Wood with Anhydrous Ammonia,* S. U. N. Y. College of Forestry, Syracuse, N. Y., 1968. 4 pp.

Davis, E. M., *Machining and Related Characteristics of United States Hardwoods,* Techn. Bull. 1267, U. S. Forest Service, Washington, D. C., 1962. 68 pp.

Farmer, R. H., *Handbook of Hardwoods,* 2nd ed., Her Majesty's Stationary Office, London, England. 243 pp.

Feist, W. C., *Finishing Wood for Exterior Applications—Paints, Stains, and Pretreatments,* U. S. Forest Products Laboratory, Madison, Wisc., 1977. 7 pp.

Feist, W. C., and E. A. Mraz, *Wood Finishing—Water Repellants and Water-Repellant Preservatives,* FPL-0124, U. S. Forest Products Laboratory, Madison, Wisc., 1978. 8 pp.

Findley, W. P. K., *Timber Pests and Diseases,* Pergamon Press, Oxford, England, 1967. 271 pp.

Forest Products Laboratory, *FPL Natural Finish,* FPL-046, U. S. F.P.L., Madison, Wisc., 1972. 7 pp.

Forest Products Laboratory, *Raised, Loosened, Torn, Chipped, and Fuzzy Grain in Lumber,* FPL-099, U. S. F.P.L., Madison, Wisc., 1965. 15 pp.

Forest Products Laboratory, *Wood—Colors and Kinds,* U.S.D.A. Handb. 101, Government Printing Office, Washington, D. C., 1956. 36 pp.

Forest Products Laboratory, *Wood Finishing—Blistering, Peeling, and Cracking of House Paint from Moisture,* FPL-0125, U. S. F.P.L., Madison, Wisc., 1970. 7 pp.

Forest Products Laboratory, *Wood Handbook—Wood as an Engineering Material,* U.S.D.A. Agr. Handb. 72, rev., Government Printing Office, Washington, D. C., 1974. 419 pp.

Forest Products Laboratory, *Wood Siding—Installing, Finishing, Maintaining,* U.S.D.A. Home and Garden Bull. 203, Government Printing Office, Washington, D. C., 1973, 13 pp.

Gillespie, R. H., D. Countryman, and R. F. Blomquist, *Adhesives in Building Construction,* U.S.D.A. Handb. 516, Government Printing Office, Washington, D. C., 1978. 160 pp.

Gjovik, L. R., and R. H. Baechler, *Selection, Production, Procurement and Use of Preservative-Treated Wood—Supplementing Federal Specification TT-W-571,* FPL-15, U. S. Forest Products Laboratory, Madison, Wisc., 1977. 37 pp.

Godshall, W. D., and J. H. Davis, *Acoustical Absorption Properties of Wood-Base Panel Materials,* FPL 104, U. S. Forest Products Laboratory, Madison, Wisc., 1969. 8 pp.

Goldstein, I. S., Degradation and Protection of Wood from Thermal Attack, in D. D. Nicholas, Ed., *Wood Deterioration and Its Prevention by Preservative Treatments,* Vol. 1, Syracuse University Press, Syracuse, N. Y., 1973, pp. 307-339.

Harkin, J. M., and J. W. Rowe, *Bark and its Possible Uses,* FPL-091, U. S. Forest Products Laboratory, Madison, Wisc., 1969. 41 pp.

Harlow, W. M., *Inside Wood—Masterpiece of Nature,* American Forestry Association, Washington, D. C., 1970. 120 pp.

Hawley, L. F., Combustion of Wood, in L. E. Wise and E. C. Jahn, Eds., *Wood Chemistry*, 2nd ed., Vol. 2, Reinhold, New York, 1952, pp. 817-825.

Hoyle, R. J., Jr., *Wood Technology in the Design of Structures,* rev. ed., Mountain Press, Missoula, Mont., 1978. 400 pp.

Jones, R. E., *Sound Insulation Evaluation of High-Performance Wood-Frame Party Partitions Under Laboratory and Field Conditions,* FPL 309, U. S. Forest Products Laboratory, Madison, Wisc., 1978. 37 pp.

Kirbach, E. D., *Methods of Improving Wear Resistance and Maintenance of Saw Teeth,* Forintek Canada Corp., Western Forest Products Laboratory Technical Report 3, Vancouver, B. C. 1979. 45 pp.

Koch, P., *Utilization of the Southern Pines,* U.S.D.A. Agr. Handb. 420, Government Printing Office, Washington, D. C., 1972. 1663 pp.

Koch, P., *Wood Machining Processes,* Ronald, New York, 1964. 530 pp.

Kollmann, F. F. P., and W. A. Côté, *Principles of Wood Science and Technology,* Vol. 1, *Solid Wood,* Springer-Verlag, Berlin, Germany, 1968. 592 pp.

Kollmann, F. F. P., E. W. Kuenzi, and A. J. Stamm, *Principles of Wood Science and Technology,* Vol. 2, *Wood Based Materials,* Springer-Verlag, Berlin, Germany, 1975, 703 pp.

Kramer, P. J., and T. T. Kozlowski, *Physiology of Woody Plants,* Academic, New York, 1979. 832 pp.

Kubler, H., Drying Tree Discs Simply and Without Defects, *Forest Prod. J.* **24**, (7), 33-35 (1974).

Kubler, H., L. Liang, and L. S. Chang, Thermal Expansion of Moist Wood, *Wood and Fiber* **5** (3), 257–267 (1973).

Kukachka, B. F., *Properties of Imported Tropical Woods,* FPL 125, U. S. Forest Products Laboratory, Madison, Wisc., 1970. 67 pp.

Laminated Timber Institute of Canada, *Timber Design Manual,* LITC, Ottawa, 1972. 458 pp.

Lunstrum, S. J., *Circular Sawmills and their Efficient Operation,* U.S.D.A. State and Private Forestry, Atlanta, 1972. 86 pp.

Lutz, J. F., *Buckle in Veneer,* FPL-0207, U. S. Forest Products Laboratory, Madison, Wisc., 1970. 11 pp.

Lutz, J. F., *Wood Veneer—Log Selection, Cutting, and Drying, U. S. Forest Service Techn. Bull. 1577,* Forest Products Laboratory, Madison, Wisc., 1978. 137 pp.

Martin, R. E., Thermal Properties of Bark, *Forest Prod. J.* **13**, (10), 419–426 (1963).

McNatt, J. D., *Basic Engineering Properties of Particleboard,* FPL 206, U. S. Forest Products Laboratory, Madison, Wisc., 1973. 14 pp.

National Bureau of Standards, *NBS Voluntary Product Standards,* Government Printing Office, Washington, D. C., various years.

National Forest Products Association, *Design Values for Wood Construction,* NFPA, 1619 Massachusetts Ave., N.W. Washington, D. C., 1979. 20 pp.

National Hardwood Lumber Association, *Rules for the Measurement and Inspection of Hardwood and Cypress Lumber,* Chicago, 1978. 115 pp.

National Paint, Varnish and Lacquer Association, *Finishing Wood Furniture,* NPVLA, Washington, D. C., 1974. 3 pp.

Nicholas, D. D., Ed., *Wood Deterioration and Its Prevention by Preservative Treatment,* Vol. 1, *Degradation and Protection of Wood,* 380 pp., Vol. 2, *Preservatives and Preservative Systems,* 402 pp., Syracuse University Press, Syracuse, N. Y., 1973.

Page, R. H., and L. Wyman, *Hickory for Charcoal and Fuel,* U. S. Forest Service SE For. Exp. Station, Asheville, N. C., 1969, 9 pp.

Panshin, A. J., and C. de Zeeuw, *Textbook of Wood Technology,* Vol. 1, *Structure, Identification, Uses, and Properties of the Commerical Woods of the United States and Canada,* 3rd ed., McGraw-Hill, New York, 1970. 705 pp.

Peck, E. C., *Bending Solid Wood to Form,* Agr. Handb. 125, U. S. Forest Service, Washington, D. C., 1957. 37 pp.

Rasmussen, E. F., *Dry Kiln Operator's Manual,* U.S.D.A. Agr. Handb. 188, Government Printing Office, Washington, D. C., 1961. 197 pp.

Rietz, R. C., and R. H. Page, *Air Drying of Lumber,* U.S.D.A. Agr. Handb. 402, Government Printing Office, Washington, D. C. 1971. 110 pp.

Scheffer, T. C., and A. F. Verrall, *Principles for Protecting Wood Buildings from Decay,* FPL 190, U. S. Forest Products Laboratory, Madison, Wisc., 1973. 57 pp.

Schultz, T. J., Acoustical Properties of Wood—a Critique of the Literature and a Survey of Practical Applications, *Forest Prod. J.* **19** (2), 21–29 (1969).

Selbo, M. L., *Adhesive Bonding of Wood,* U. S. Forest Service Techn. Bull. 1512, Government Printing Office, Washington, D. C., 1975. 122 pp.

Shelton, J. W., and A. B. Shapiro, *Woodburner's Encyclopedia,* 4th ed., Vermont Crossroad Press, Waitsfield, Vt., 1977. 156 pp.

Siau, J. F., *Flow in Wood,* Syracuse University Press, Syracuse, N. Y., 1971. 131 pp.

Skaar, C., *Water in Wood,* Syracuse University Press, Syracuse, N. Y., 1972, 218 pp.

Skeist, I., *Handbook of Adhesives,* 2nd ed., Van Nostrand Reinhold, New York, 1977. 921 pp.

Smith, W. S., *Simplified Guidelines to Hardwood Lumber Grading,* U. S. Forest Service SE For. Exp. Station, Asheville, N. C., 1967. 24 pp.

Stamm, A. J., *Wood and Cellulose Science,* Ronald, New York, 1964. 549 pp.

Stevens, W. C., and N. Turner, *Wood Bending Handbook,* Ministry of Technology, London, England, 1971. 200 pp.

Stumbo, D. A., Influence of Surface Aging Prior to Gluing on Bond Strength of Douglas-Fir and Redwood, *Forest Prod. J.* **14** (7), 582–589 (1964).

Suchsland, O., Selecting Panel Materials by Furniture and Cabinet Manufacturers, *Res. Bull.* **27**, Michigan State University, East Lansing, Mich., 1970. 32 pp.

Thompson, W. A., *Permeability and Weathering Properties of Film-Forming Materials Used in Wood Finishing,* Res. Rep. 4, Miss. State Univ., State College, Miss., 1968. 32 pp.

Western Wood Products Association, *Western Woods Use Book—Structural Data and Design Tables*, WWPA, Portland, Or., 1979. 310 pp.

Appendix

Wood Identification

Here you learn how to recognize 30 wood species commonly used in North America. Some of them can be determined at a cursory glance on basis of figure and color, while the others require close inspection and knowledge of wood structure as provided in Sections 1.2, 1.3, and under *resin* in Section 2.1. Identification of wood species that are not covered here follows the same principle, that is, comparing observed structural characteristics with species descriptions, which in this case would have to be looked up in the literature.

TOOLS NEEDED AND SAMPLE PREPARATION

For close inspection you need a 10× hand lens and a utility knife with razor blade, preferably one of the heavy-duty retractable blade type. The knife is to smoothen the wood surface and to uncover the species characteristics. Sanding is inferior to cutting because it clogs pores and produces scratches.

Most cuts have to be made at rough end-grain surfaces, which harbor the majority of the species characteristics. To facilitate this relatively difficult transverse cutting, you may first want to soften the spot by moistening. Then press the piece firmly against the chest, or hold it between the knees, to cut an area of at least one annual ring, pushing the knife forward in one fast move and peeling off a thin unbroken slice to obtain a continuous surface. Chipping off small fractions in a series of small jabs produces surfaces too rough for the examination. Species with hard latewood bands cut best along the annual rings, while others work better along the rays. Apply the same sort of dragging motion as when cutting rough meat, pushing the knife forward perpendicular to its edge and pulling simultaneously in the direction of the knife-edge. Oblique cuts are relatively easy but reveal less than exact cross sections.

Clamping the sample into a workbench or a vice facilitates cutting in fast continuous moves. Going one step farther, skilled woodworkers use a razor-sharp hand plane or a spokeshave in place of the utility knife. By far superior to any other equipment is a microtome, which cuts slices of adjustable thickness from clamped cubes which have been softened by boiling in water.

Side grain surfaces of a sample, by contrast, need to be about as large as the palm of your hand. Planed or sanded they rarely need cutting, except to expose rays at exact radial surfaces. Splitting instead of cutting produces deficient uneven surfaces.

The lens reveals many characteristics which are too small for detection by the unaided eye. Hold it close to the eye and bring the wood surface within focus without shutting out the light. Wetting accentuates many minute features and wood figure characteristics, particularly the earlywood-latewood contrast.

HOW TO DISTINGUISH BETWEEN HARDWOODS AND SOFTWOODS

When confronted with an unknown piece, first determine whether it is hardwood or softwood. Hardwoods have vessels, softwoods have not. In cross section, vessel pores are clearly visible at least under the lens, while the cavities of softwood tracheids are too small in most pieces to be seen with a 10× lens. The tracheids comprise about 90% of the cross section and form radial files, whereas vessel pores occupy at most two thirds and are scattered among fibers (Fig. 1.3).

Beginners may mistake resin canals of softwoods for hardwood vessels. The two are different in number and distribution, the vessels being much more numerous and more regular in occurrence. Resin canals and vessels cannot be distinguished by size; both vary in this regard from species to species in similar ranges. The size of resin canals is fairly constant within the species, but so is that of vessels in several hardwoods. Openness cannot serve as a criterion either, because vessels as well as canals may be plugged with tyloses and resin, respectively. At side-grain surfaces, pores form narrow grooves and contribute to the figure in many species, whereas the resin canals show as a few lines which because of dust sticking to the resin tend to be dark and not sharply defined. Length of vessel grooves and of canal lines depends on the orientation of the surface plane with regard to the fiber direction; where the plane exactly follows the fiber direction, both, grooves and lines, extend over the whole length of the piece.

STYLE OF THE KEYS

The species characteristics are listed in two keys, one for hardwoods and one for softwoods. The surface to be observed is designated with the symbol (*x*) for *cross section*, (*s*) for *side-grain surface*, (*t*) for *tangential surface*, and (*r*) for *radial surface*. Capital letters (*X, S, T, R*) mean *observation under the lens.* No surface designation is given for characteristics that are visible on all surfaces without lens. Cardinal species characteristics appear in *italics.* The numbers stand for the average specific gravity, based on weight when ovendry and volume at 12% moisture content; because of variations within each species, they serve only as a crude identification aid.

When observing color and tyloses, keep in mind that the keys concern heartwood only. Sapwood is yellowish or whitish in all species and not well suited for wood identification. Observe color at the side grain, considering that it varies even within heartwood of individual trees and changes during drying as well as under exposure to light; therefore fresh-cut surfaces are preferable.

In several instances the keys point to a genus or to several species within a genus, rather than to one species, the reason being that the species in that group do not differ under the lens. Some, such as the group southern yellow pine, look identical even under the microscope, so that for their identification into species even expert wood anatomists have to resort to the tree itself. I give the *commercial names for lumber* as listed in the *American Softwood Lumber Standard PS 20–70* and in the *Nomenclature of Domestic Hardwoods and Softwoods ASTM D 1165* for hardwoods. *Khaya* is also known as *African mahogany, lauan* as *Phillippine mahogany.* For tree names see Tables 1.1 and 4.2.

Some woods resemble others that are not included, making the identification somewhat uncertain. Softwoods are generally more difficult to recognize than the more diversified hardwoods; many softwood differences have to be described in terms of subtleties, whose detection requires not only knowledge but also a perceptive eye. Since descriptions cannot be as close and concise as samples, keep known samples on hand for comparison.

At this point it may be of interest that the U.S. Forest Products Laboratory (P.O. Box 5130, Madison, Wis. 53705) maintains a free wood identification service. The FPL's Center for Wood Anatomy Research employs a wide variety of tools and can identify with a very high degree of certainty. Submitted samples may be as small as slivers, but larger samples, preferably large enough to be easily hand-held for cutting, are always better.

FEATURES MENTIONED IN THE KEYS

The terms *ring-porous* and *diffuse-porous* are explained in Section 1.3. In the *semi-ring-porous* walnut the pores are slightly graded in size from earlywood to latewood. Tyloses, as thin membranes of froth- or balloon-like growths in vessel pores, may be confused with sanding dust (Figs. 1.17 and 1.18); when in doubt, cut off a few cross-section slices—superficial dust disappears whereas tyloses which occupy the whole vessel remain. Gum deposits fill a fraction of the length of some vessels and therefore appear clearest at side grain surfaces, looking like dark tiny grains, as shown for khaya.

Rays resemble knife blades inserted radially between fibers, appearing as lines at cross sections, as spindles at tangential sections, and as bands at radial sections. (Fig. 1.5). *Ray height* means *the length of the spindles.* At intermediate side grain surfaces the rays look like thick spindles, in nearly radial sections having the shape of irregular patches (Fig. 1.6).

Rays consist mainly of parenchyma tissue, which in color and light reflection differs from fiber tissue. Parenchyma that surrounds pores, as in ash, makes the pores at the cross section look like white dots. Other parenchyma cells form narrow tangential bands that appear as fine broken lines within the annual ring in hickory, and as continuous *marginal* lines at the end of the annual ring in many other species. In the tropical woods mahogany and khaya these continuous lines create the impression of annual rings in otherwise uniform wood.

Tangentially split ponderosa pine has scattered tiny depressions (dimples) at the pith side surface and corresponding pimples at the bark side surface. The dimples and pimples result from local undulations of fibers, as if during wood growth grains of sand under the bark had indented the cambial sheath. Planed surfaces naturally lack the depressions and pimples, but the undulations influence reflection of light and cause a dimpled appearance at certain angles of light. In *bird's eye* grain, occurring occasionally in hard maple, the undulations are closer together and clearly visible in any angle of light.

The photographs are supposed to show the wood as seen by a reader with no tools other than a utility knife and a 10× lens. A few microscopic cross sections of greater technical complexity are included for the sake of clarity, particularly the photograph of basswood, made from a thin slice illuminated from beneath.

HARDWOOD KEY

I Pores conspicuous, visible at all surfaces without lens.

A Ring-porous.

1 Rays large and distinct, at least 1/2 in. high (*t*).

(a) Rays about 1/2 in. high (*t*, Fig. A.1*a*). Wood reddish or pinkish. Tyloses absent (*XS*, Fig. A.2*a* and *c*). 0.63 **Red oak**

(b) Rays 1 in. and higher (*t*, Fig. A.1*b*). Wood light brown or buff. Earlywood pores generally plugged with tyloses (*XS*, Fig. A.2*b* and *d*). 0.65. **White oak**

2 Rays small, less than 1/16 in. high, visible or barely visible (*t*), at exactly radial surface clearly visible at certain angle of light (*r*).

(a) Latewood pores arranged in *wavy tangential light-colored lines* (*x*, Fig. A.3*a*), causing feathered pattern (*t*, Fig. A.3*b*). 0.52 **Soft elm**

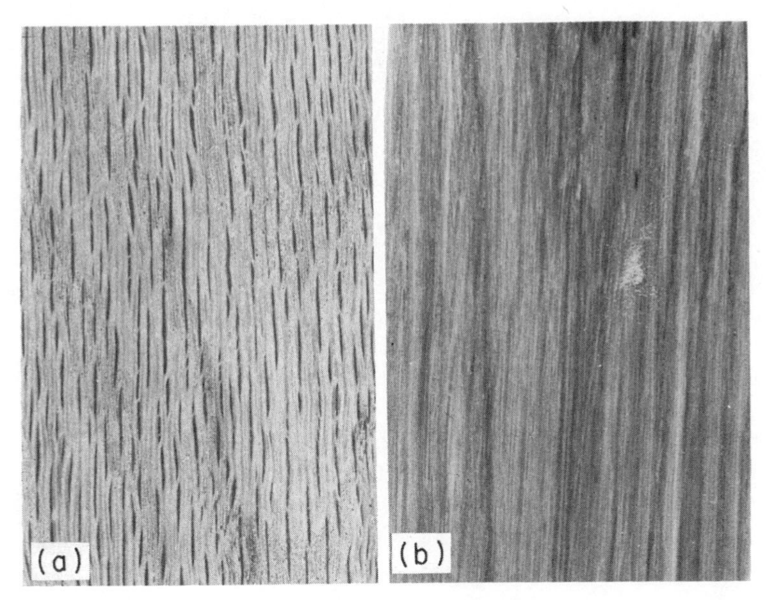

Figure A.1. Tangential surfaces of red oak (*a*) and white oak (*b*), showing different height of rays.

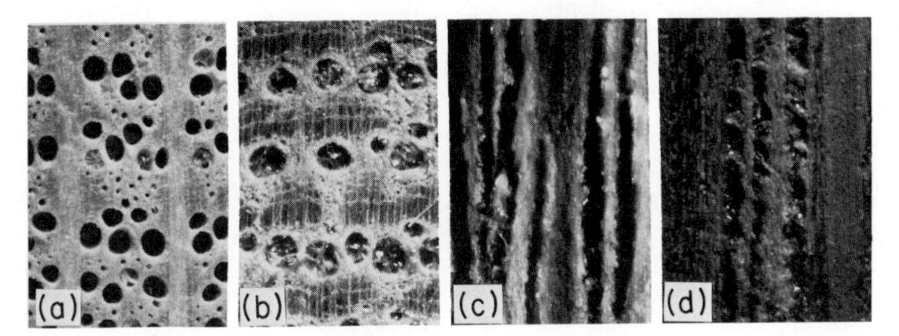

Figure A.2. Cross sections (*a*, *b*) and lateral sections (*c*, *d*) of red oak (*a*, *c*) and white oak (*b*, *d*); ×8.

(b) Latewood pores appear as *light-colored dots* (*X*, Fig. A.3*c*). Short tangential bands of parenchyma near the end of the annual ring (*X*). Occasionally feathered pattern near the end of the ring (*t*). 0.58. **Ash**

(c) *Fine tangential lines* of parenchyma in latewood (*X*). Transition from earlywood to latewood gradual, creating the impression of semi-ring porous wood (*X*). Pores in earlywood irregularly and widely spaced, usually in one single row (*X*, Fig. A.3*d*). 0.7. **Hickory**

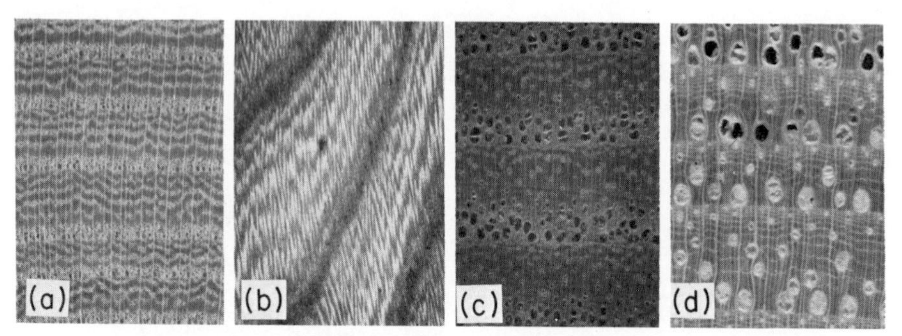

Figure A.3. Cross sections of soft elm (*a*), ash (*c*) and hickory (*d*); ×6. (*b*) Tangential surface of soft elm at an angle of light that brings out the feathered pattern.

(d) Wood distinctive yellow-brown, often with dark streaks, greasy feel. Odor of old leather. Tyloses and solid deposits in vessels (*XS*). 0.63. **Teak**

B Diffuse- and semi-ring porous.

1 Rings distinct, rather regularly spaced, visible at least under certain angle of light.

(a) Semi-ring-porous (Fig. A.4*a*). Wood chocolate to purplish brown. Distinct odor. 0.55 **Walnut**

(b) Pores uniform in size throughout the ring (*X*). Tangential rings formed by continuous light-colored lines, 1/32 to 1/2 in. apart (*xr*, Fig. A.4*b* and *c*), visible under certain angle of light (*s*, Fig. A.4*c*). Wood tan to golden brown. Dark amber-colored gum in pores (*S*). 0.5. **Mahogany**

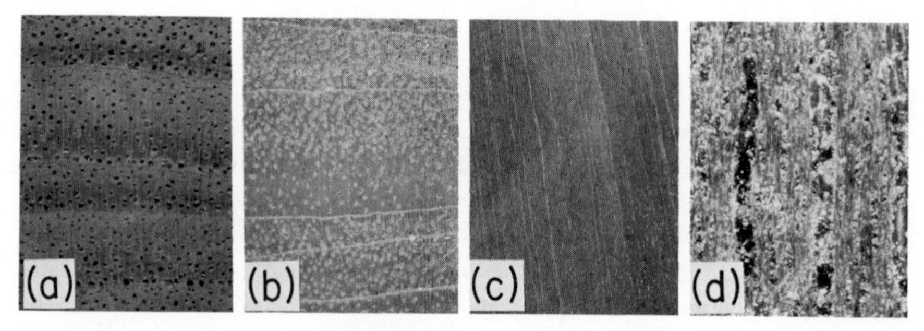

Figure A.4. (*a*) Cross section of walnut ×3. (*b*) Cross section of mahogany ×2. (*c*) Radial section of mahogany ×1/2. (*d*) Lateral section of khaya ×14, showing gum deposits in the vessel at left.

2 No distinct rings, but sometimes ribbons formed by interlocked grain (*r*).

(a) Rare light-brown tangential lines, irregularly spaced and barely visible (*x*). Wood lustrous-brown to reddish-brown. Dark gum in pores (*S*, Fig. A.4*d*). 0.52. **Khaya**

(b) *Pores very large* (*x*). *No dark gum* but partly tyloses in pores (*S*). Tangential lines 1/8 in to several inches apart, frequently discontinuous (*x*), appearing as rows of minute openings filled with a white substance (*X*). 0.45. **Lauan**

II Pores inconspicuous, not visible (*x*), diffuse-porous (*XS*s).

A Rays conspicuous, easily found on all surfaces, about 1/16 to 1/4 in. high (*t*).

1 Numerous broad rays, occupying about 1/3 of the surface area, closely and evenly spaced (*t*); their color contrasts with fiber tissue (*t*, Fig. A.5*a*). 0.49. **Sycamore**

2 Few broad rays (*tx*), but numerous fine rays that do not sharply contrast with fiber tissue (*t*), irregularly spaced (*X*, Fig. A.5*b*). 0.64. **Beech**

B Rays small, less than 1/16 in. high (*tT*).

1 Rings (*t*) and rays (*r*) visible.

(a) Pores still visible (*s*). Rays invisible (*t*), occupy less than half the surface area (*X*).

α Wood reddish-brown. Rays barely visible, are numerous and close together but narrow (*x*). *Pores numerous, crowded* especially at the beginning of each ring (*X*, Fig. A.6*a*). 0.50. **Cherry**

β Wood light to reddish-brown. Rays not visible. *Pores not crowded* (*X*, Fig. A.6*b*). Rings frequently not clearly defined. 0.62. **Birch**

(b) Pores barely visible (*s*), appear numerous and crowded. Wood yellowish brown with a greenish tinge. Rings sometimes almost invisible (*s*), clearly defined by a *fine line of marginal parenchyma* (*X*, Fig. A.7*a*). Rays invisible (*t*), very close, occupy half the area, uniform in width (*X*). 0.42. **Yellow poplar**

(c) Pores not visible (*s*), not crowded (*X*, Fig. A.6*c*). Rays barely visible (*t*), distinct (*r*). Rings clearly defined by a *reddish-brown layer* (*s*). 0.63. **Hard maple**

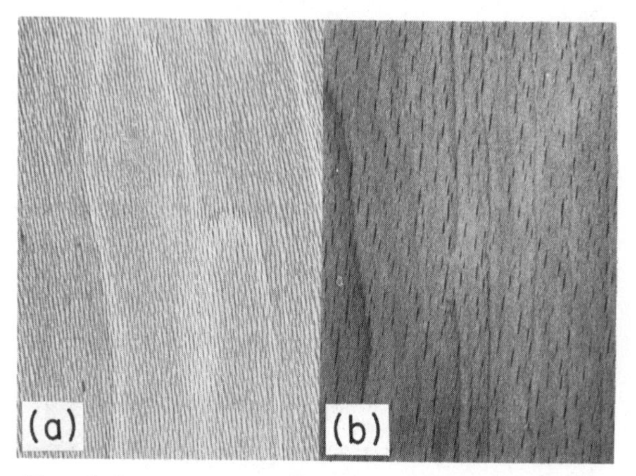

Figure A.5. Tangential surface of sycamore (*a*) and beech (*b*).

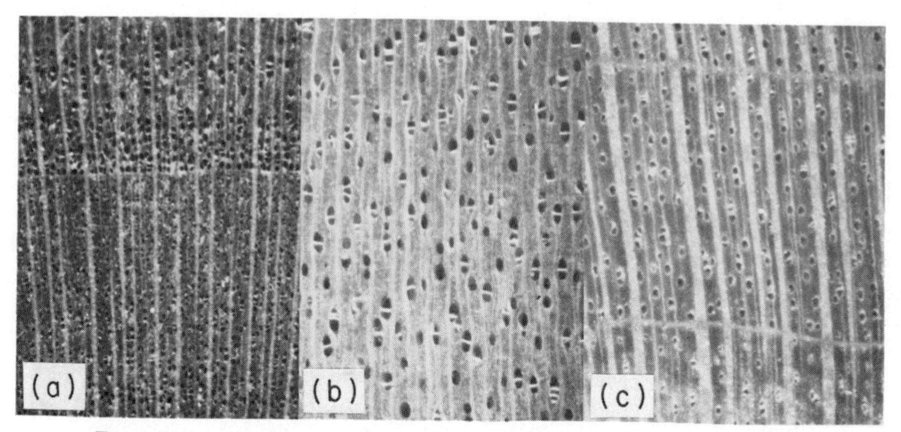

Figure A.6. Cross sections of cherry (*a*), birch (*b*), and hard maple (*c*); ×15.

2 Little or no figure (*t*).

(a) Wood pink to deep brown in streaks. Rings not discernible (*x*), generally even not with lens (*X*). Very fine rays occupy almost half the area (*X*), visible (*r*). Pores barely visible (*s*). 0.52. **Gum**

(b) Heartwood creamy white to pale creamy brown, not distinctly darker than sapwood. Rays occupy much less than half the area (*X*). Pores not visible (*s*).

α Wood light buff, sometimes with a pinkish cast. Rays visible and of same color as background (*r*), uniformly

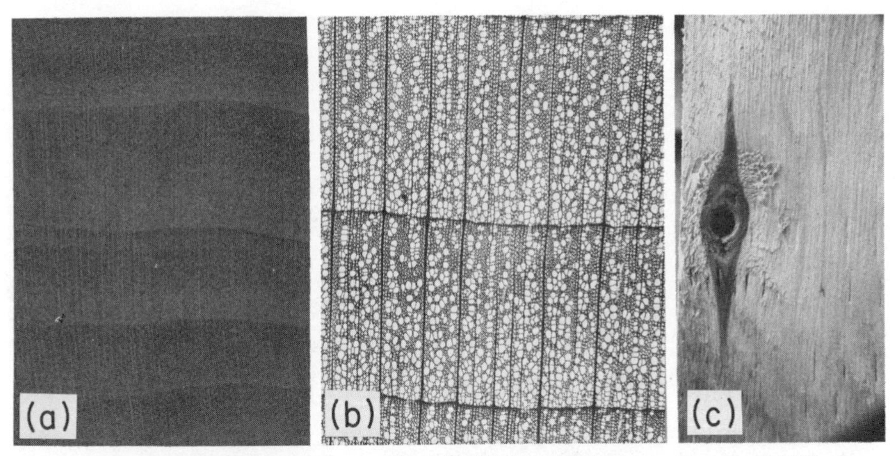

Figure A.7. (*a*) Cross section of yellow poplar ×3. (*b*) Microphotograph of basswood ×15, showing lace-like pattern of rays. (*c*) Radial surface of aspen × 1/2, showing discoloration above and below knot.

spaced and seemingly discontinuous, forming a *lace-like pattern* (*X*, Fig. A.7*b*). Faint odor when wetted. 0.37. **Basswood**

β Wood almost white, sometimes with brown areas especially around knots (Fig. A.7*c*). *Rays barely visible* or indistinct (*X*), not visible (*r*). Pores gradually decreasing size from earlywood to latewood (*X*). 0.38. **Aspen**

SOFTWOOD KEY

I Resin canals present (*Xs*).

A Resin canals large and conspicuous (*Xxs*), present in most annual rings (*Xx*). Resinous *pine* odor.

1 Rings indistinct, gradual transition (Fig. A.8*a* and *d*). 0.35. **Northern white pine**

2 Rings distinct, abrupt transition.

(a) Latewood bands narrow, comprising about 1/5 of the ring (Fig. A.8*b* and *e*). Dimples frequent (*t*, Fig. A.8*e*). 0.4. **Ponderosa pine**

(b) Latewood bands wide, usually much wider than 1/5 of the ring (Fig. A.8*c* and *f*). Wood coarse. 0.55. **Southern yellow pine**

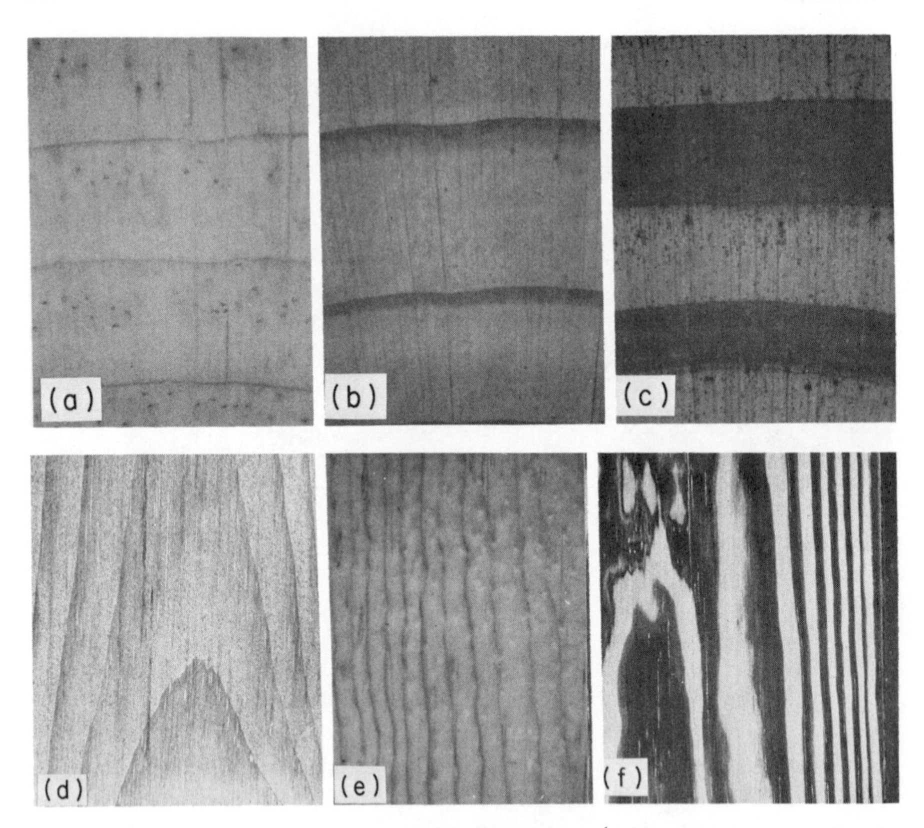

Figure A.8. Cross sections ×4 (*a*, *b*, *c*) and lateral surfaces × 1/2 (*d*, *e*, *f*) of northern white pine (*a*, *d*), ponderosa pine (*b*, *e*), and southern yellow pine (*c*, *f*). (*e*) Shows dimples.

B Resin canals small and sparse, absent in some rings (*X*).

1 Rings indistinct, earlywood and latewood rather uniform in color. Wood creamy white with yellowish cast, lustrous, may be dimpled (*t*). 0.4. **Spruce**

2 Rings distinct, bands of latewood prominent, abrupt transition.

(a) Resin canals frequently in short tangential rows in latewood or near latewood (*X*, Fig. A.9*a*), found with little difficulty (*s*). Wood pinkish or reddish. Characteristic odor. 0.48. **Douglas fir**

(b) Resin canals difficult to find, sometimes in tangential pairs or in clusters of 3 to 5 (*X*, Fig. A.9*b*). Wood brownish rather than reddish. No characteristic odor. 0.52. **Western larch**

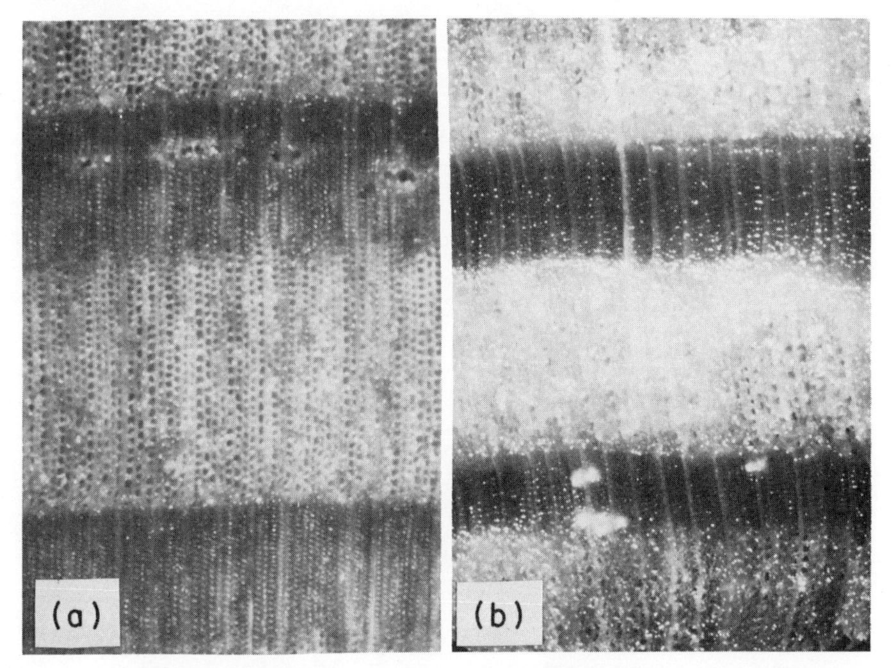

Figure A.9. Wood cross sections showing resin canals in a tangential row, in Douglas fir (*a*), and in a cluster, in western larch (*b*); ×25.

II No resin canals.

A Wood light, pale or slightly pinkish or with purplish tint, no sharp contrast between heartwood and sapwood.

1 Earlywood and latewood contrast in color; latewood appears light brown or tan and earlywood almost white (*s*). 0.39. **White fir**

2 Earlywood and latewood show little contrast in color (*s*). Purplish tinge. 0.45. **West coast hemlock**

B Wood dark or brownish red.

1 Aromatic *shingle* odor. 0.32. **Western red cedar**

2 No aromatic odor. 0.40. **Redwood**

3 Unpleasant, rancid odor. Oily or greasy surface (*s*). Abrupt transition. Color variable. 0.46. **Cypress**

Index